Advances in
Neural Signal Processing

Edited by Ramana Vinjamuri

Published in London, United Kingdom

IntechOpen

Supporting open minds since 2005

Advances in Neural Signal Processing
http://dx.doi.org/10.5772/intechopen.81424
Edited by Ramana Vinjamuri

Contributors
Tal Dotan Ben-Soussan, Fabio Marson, Patrizio Paoletti, Stefano Lasaponara, Joseph Glicksohn, Antonio De Fano, Dimiter Prodanov, Shyam Diwakar, Chaitanya Nutakki, Sandeep Bodda, Vito Di Maio, Silvia Santillo, Rajesh Reghunadhan, Fasil O.K., T. Cetin Akinci, Serhat Seker, Deniz Türkpençe, Ufuk Korkmaz, Furkan Kalin, Emília Quinta-Ferreira, Carlos Matias, Johnattan C.S. Freitas, Fernando D.S, Sampaio Dos Aidos, Paulo J. Mendes, José C. Dionísio, Rosa M. Santos, Luís M. Rosário, Rosa M. Quinta-Ferreira, João N. Miraldo, Ramana Vinjamuri, Dingyi Pei

First published in London, United Kingdom, 2020 by IntechOpen
IntechOpen is the global imprint of INTECHOPEN LIMITED, registered in England and Wales, registration number: 11086078, 7th floor, 10 Lower Thames Street, London, EC3R 6AF, United Kingdom
Printed in Croatia

British Library Cataloguing-in-Publication Data
A catalogue record for this book is available from the British Library

Additional hard and PDF copies can be obtained from orders@intechopen.com

Advances in Neural Signal Processing
Edited by Ramana Vinjamuri
p. cm.
Print ISBN 978-1-78984-113-8
Online ISBN 978-1-78984-114-5
eBook (PDF) ISBN 978-1-83968-396-1

An electronic version of this book is freely available, thanks to the support of libraries working with Knowledge Unlatched. KU is a collaborative initiative designed to make high quality books Open Access for the public good. More information about the initiative and links to the Open Access version can be found at www.knowledgeunlatched.org

For EU product safety concerns:
IN TECH d.o.o., Prolaz Marije Krucifikse Kozulić 3, 51000 Rijeka, Croatia, info@intechopen.com or visit our website at intechopen.com.

Meet the editor

Ramana Vinjamuri received his undergraduate degree in Electrical Engineering from Kakatiya University (India) in 2002. He received his MS in Electrical Engineering from Villanova University in 2004 with a specialization in Bioinstrumentation. He received his PhD in Electrical Engineering from the University of Pittsburgh in 2008 with a specialization in Dimensionality Reduction in Control and Coordination of the Human Hand. He worked as a postdoctoral fellow (2008–2012) in the field of Brain Machine Interfaces (BMI) to control prostheses in the School of Medicine, University of Pittsburgh. He worked as a Research Assistant Professor in the Department of Biomedical Engineering at Johns Hopkins University (2012–2013). He is currently a Harvey N Davis Distinguished Assistant Professor in the Department of Biomedical Engineering at Stevens Institute of Technology. Dr. Vinjamuri received the NSF CAREER Award in 2019. He also holds a secondary appointment as an Adjunct Assistant Professor at the Indian Institute of Technology, Hyderabad, India.

Contents

Preface

In September 2018, I was approached by IntechOpen to see if I was interested in serving as editor for this project, *Advances in Neural Signal Processing*. Neural signal processing is a specialized area of signal processing that is aimed at extracting information or decoding intent from neural signals recorded from the central or peripheral nervous system. This has significant applications in the areas of neuroscience and neural engineering. These applications are famously known in the area of brain–machine interfaces. This book presents recent advances in this flourishing field of neural signal processing with demonstrative applications.

In this book, readers will find chapters on three major applications of neural signal processing methodologies in basic science research and neuroprosthetic and clinical applications. Basic science research includes modeling zinc dynamics in hippocampal mossy fiber synaptic cleft and information processing and synaptic transmission. Neuroprosthetic applications consist of decoding upper limb movements and gait phase kinematics using electroencephalography. Clinical applications include detection, prediction, and classification of seizures.

I would like to thank the authors for their valuable contributions, especially my students from the Sensorimotor Control Laboratory at Stevens Institute of Technology who contributed to the introductory chapter. I would also like to thank IntechOpen staff especially Dajana Pemac, Romina Skomersic, and Andrea Koric for their kind assistance throughout the editing process. Without their help, this book would not have manifested. I am sure that interested readers will find this volume informative and inspiring.

Ramana Vinjamuri
Assistant Professor,
Director of Sensorimotor Control Laboratory,
Department of Biomedical Engineering,
Stevens Institute of Technology,
New Jersey, USA

Chapter 1

Introductory Chapter: Methods and Applications of Neural Signal Processing

Dingyi Pei and Ramana Vinjamuri

1. Introduction

Analytical methods are crucial to advance the field of brain sciences, and efficient and effective methods of data analysis are required. Early from the last century, the neural signals have been used in the engineering sphere to discover mechanisms by which neural activity is generated and corresponding behavior is produced. The function of the neural system was detected and studied using engineering methodologies, and meanwhile, the engineering methodologies helped to understand, repair, replace, enhance, or otherwise use the properties and functions of neural systems. The neural signals are recorded by advanced neural recording technologies, and the information is extracted to be used for the understanding of neural representations of behavior. The external devices are designed to assist signal acquisition, signal processing, or provide neural feedback to humans.

Since movement is an essential activity of daily life, some of the major applications of neural engineering in the field of motor control typically involve motor function compensation, movement restoration, rehabilitation, disorder detection, etc. A movement process is integrated and translated from the higher levels of the control system, and it involves a series of transmissions to multi-structure musculoskeletal coordination. The central nervous system (CNS) works as a computational controller structure in motor behavior characterization and reorganization [1]. Multiple structures in the brain contribute to motor control by connecting, integrating, and coordinating the motor-related information. Each structure is utilized in formulating a motor command when a particular action is performed, and the CNS switches the command between multiple motor-related structures [2]. The mechanism of coordination and cooperation of these structures in the brain could be determined as "black box" models, providing the neural representations of relationships between motor command input and predicted behavior output. These models may represent multiple brain structures, especially the regions with synaptic plasticity that can receive and send out information.

2. Neural recording and stimulation

Populations of neurons exhibit time-varying fluctuations in their aggregate activity. Currently, various invasive or noninvasive recordings exist that can record large amounts of spatial and temporal information from the human and nonhuman

IntechOpen

brain. In order to investigate how the motor-related information is generated, and what kind of patterns could be found in certain areas during a specific action, many engineering methodologies are applied.

In the human brain, neurons communicate with each other through connections known as synapses. Synapses can be electrical or chemical, and the excitatory or inhibitory nature of synapses contributes to information transmission—the influx and outflux of sodium and potassium causing the membrane potential to rise and fall rapidly. The rapid changes of membrane potentials are called spikes, which can be recorded by intercellular or extracellular recordings. Valuable information can be discovered from the rate of spikes, namely, the firing rate. The deep brain implanted electrodes allow the recording from individual neurons and can present significant results in awake animals but not in humans. Multielectrode arrays can record the voltage oscillations from multiple neurons. Simultaneously recording from a large population of local neurons increases spatial resolution benefits to the extraction of complex information in contrast with single-unit recordings. The aforementioned invasive recording technologies provide considerably less vulnerability to artifacts and relevantly higher resolution and larger amplitudes (voltages), and thereby the performance relies much more on the technologies of electrodes. However, there are several limitations of these invasive recording technologies including restricted to clinical environments and the risks of surgery and implantations.

As an alternative to the constrained invasive technologies, several noninvasive recording technologies such as electroencephalography and magnetoencephalography have been used in human studies. Advanced computational algorithms promise to promote signal processing and signal filtering; thus, more and more noninvasive recording technologies are being considered in human studies. Some techniques record neuronal potentials from the scalp, and such recordings capture the population activity of thousands of neurons depending on the level of recording. Multiple layers restrict information transmission from the cerebral cortex to the scalp leading to lower amplitudes of the signal and lower spatial resolution. Additionally, the electrodes are sensitive to the surrounding interferences like eye movements, facial movements, chewing, swallowing, etc. Therefore, it is necessary to apply robust and efficient signal processing technologies to amplify the neural activity and filter out the ambient and transducer noise, thus improving the signal-to-noise ratio.

Under noninvasive technologies, there are imaging methods that focus on the metabolic activity in the brain rather than the activity of neurons or the population of neurons. When performing a specific task, the activation of the brain neurons is enhanced and thereby more oxygen is required and absorbed from surrounded blood vessels. An increased inflow and higher oxygenated level can be detected. This hemodynamic response is comparatively slow that it reaches the peak in a few seconds and takes a longer time to fall back to the original level. Therefore, this kind of recording technology provides good spatial resolution but very poor temporal resolution.

In addition to neural recording technologies, there are also neural stimulation technologies that are used in clinical treatments (cochlear implants and deep brain stimulators) and emerging neuroprosthetics. This involves giving electrical or magnetic stimulation to a particular region of the brain to mimic sensorimotor feedback. Most recording electrodes can also be used for stimulations. Brain stimulations have proven effective in clinical treatments. These methodologies also involve the use of signal processing methodologies in determining ideal stimulation patterns. **Table 1** summarizes neural recording and stimulation technologies.

Electrical recordings	Single-unit recordings (spikes)	Microelectrodes insert into neurons or placed between adjacent neurons
	Local field potential (LFP) recordings	Multielectrode arrays placed inside the brain
	Electrocorticography (ECoG)	Implanted electrodes placed on the upper layers of cerebral cortex
	Electroencephalography (EEG)	Electrodes placed on the surface of the scalp
Magnetic recordings	Magnetoencephalography (MEG)	Measures the magnetic field produced by electrical activity in the brain
Neuroimaging recordings	Functional near-infrared recordings (fNIR)	Detects near-infrared light absorbance of hemoglobin in the blood with/without oxygen
	Functional magnetic resonance imaging (fMRI)	Measures the changes in oxygenated and deoxygenated hemoglobin concentrations in the blood
	Positron emission tomography (PET)	Detects the radioactive compound as a result of metabolic activity caused by brain activity
Brain stimulations	Transcranial magnetic stimulation (TMS)	Current-passed coil of wire paced next to the skull to produce a rapidly change magnetic field
	Transcranial direct current stimulation (tDCS)	Stimulates specific parts of the brain using low-intensity direct electrical currents
	Deep brain stimulation (DBS)	Electrodes are implanted in target regions of the brain

Table 1.
Neural recording and stimulation technologies.

3. Neural signal processing

3.1 Spike sorting

Spikes from closer neurons produce larger amplitude deflections in the recorded signal. The goal of signal processing methods for such an input signal is to reliably isolate and extract the spikes being emitted by a single neuron per recording electrode. This procedure is usually called spike sorting. The simplest spike sorting method is to classify spikes according to their peak amplitude. Sometimes, the peak amplitudes may be the same for different neurons, making the method not feasible. A better approach is the window discriminator method in which the experimenter visually examines the data and places windows on aligned recordings of spikes of the same shape. The recent trend has been toward clustering spikes automatically into groups based on shape, where each group corresponds to spikes from one neuron. The shape of a spike is characterized by features extracted using wavelets or dimensionality reduction techniques.

3.2 Temporal and spatial feature extraction

Key temporal and spatial features can represent and help us understand the neural activity from the underlying oscillations. The neural signals recorded from the brain are typically mixture potentials resulting from network activity of a large population of neurons around the local neighborhood. Thus, applying appropriate feature extraction methods can isolate and extract significant features in both temporal and spatial domains.

3.2.1 Spatial filtering

For some methods that record brain signals using multi-electrodes, the signals are recorded from multiple regions of the brain. With the large variance of global noise, the local signals appear diminished. Therefore, spatial filtering or re-referencing methods are applied to enhance the local activity and filter out the common noise. For individual electrodes, the averaged activity from surrounded electrodes (Laplacian filtering) or from global electrodes (common averaged referencing) is subtracted. Spatial filtering methods can also be used to estimate the variance of the neural data.

3.2.2 Temporal analysis

The quality of the recorded brain signals primarily depends on recording techniques. However, the recorded time-series signals contain lots of noise that can be filtered using time-domain filtering methods. Numerous filtering techniques like moving average smoothing, exponential smoothing, etc. are used to preprocess raw signals in the time domain.

In addition to filtering, temporal analysis can also be used to extract significant features that represent behavior. These significant features can be extracted out from a series of time signals using computational models. Some neural signals tend to be correlated over time, and thus, the following time samples are possible to be predicted based on the previous samples using autoregressive models (for stationary signals) or adaptive autoregressive models (for nonstationary signals). Such methods depend on the model built up from the characteristic internal relationships between the previous signal samples and the subsequent samples. The coefficients of the model can be considered as neural features for the subsequent pattern recognition or classification procedure utilized for real-time decoding or estimation.

3.2.3 Frequency analysis

While temporal analysis methods are useful, there are some signals for which these methods may not result in extracting meaningful features. For example, noninvasive methods such as EEG are based on signals that reflect the activity of several thousands of neurons. Poor spatial and temporal resolution challenges the feature extraction in the time domain. The recorded signal thus can capture only the correlated activities of large populations of neurons, such as oscillatory activity.

The intrinsic property of the brain signals is neuronal oscillations [3]. Theoretically, these oscillations can be decomposed with a set of basis functions, such as sinusoid functions using Fourier transform (FT) for periodic signals. For each cycle, the amplitude, the period, and the waveform symmetry are measured and oscillatory bursts are algorithmically identified, allowing us to investigate the variability of oscillatory features within and between bursts. Usually, for neural signals, short-time Fourier transform (STFT) provides better results by performing FT with sliding short-time windows. For the nonperiodic signals, the wavelet transform is applied for signal decomposition. A variety of scaled and finite-length waveforms can be selected according to the shape of the raw neural signals. The wavelet coefficients sometimes contain unique information which can be considered as neural features. Additionally, the power spectrum of neural signals usually reflects lots is important information, such as power spectral density (PSD).

3.2.4 Time-frequency analysis

By combining the advantages of temporal and frequency analyses, the researchers realized the power of time-frequency analysis. As an example, using decomposition techniques, a signal can be decomposed into intrinsic mode functions (IMF) and instantaneous frequencies over time can be obtained by applying methods such as Hilbert spectral analysis. The most significant advantage of this technique is that the nonlinear, nonstationary recorded neural signals can be transformed into linear and stationary components. These components are usually physically meaningful since the special features are localized in their instantaneous frequencies and represent meaningful behavioral information in the time-frequency domain. Time-frequency analysis is extensively implemented in neural signal processing since the individual analysis in the time or frequency domain comes with respective disadvantages, and time-frequency analysis trades off time and frequency resolution to get the best representation of the signals. Other techniques such as spectrogram and STFT are most performed by segmenting a signal into short periods and estimating the spectrum over sliding windows.

3.3 Dimensionality reduction

A critical procedure in neural signal processing is to reduce the high dimensionality of the recorded neural data. These data could be brain images, multi-electrode signals, network potentials, or high-dimensional neural features. Several algorithms can be applied linearly or nonlinearly to preserve the most useful components and remove redundancies. Principal component analysis (PCA) is to find the direction of maximum variance and thereby build the principal components (weighted linear combinations) based on the observed variance. Linear discriminant analysis (LDA) performs similar to PCA but tends to minimize the variance within a group of neural data and maximize the distance between groups of neural data. Thus, PCA is described as an unsupervised algorithm implemented for feature extraction, and LDA is described as a supervised algorithm that uses training based on labels for groups of data. Other methods that are used most often are CCA and ICA. Canonical correlation analysis (CCA) is yet another method for exploring the relationships between two multivariate sets of variables allowing us to summarize the relationships into a lesser number of variables while preserving the essential features of the relationships. Independent component analysis (ICA) is a blind source separation method rather than a dimensionality reduction method. Neural signals consist of recordings of potentials that are presumably generated by mixing some underlying components of brain activity. ICA can theoretically isolate these underlying components of brain activity by computing independent components. Additionally, ICA can also be used as a filtering method to remove signal artifacts generated by eye blinks or other artifacts in EEG signals.

3.4 Machine learning algorithms

Machine learning or deep learning algorithms have become increasingly popular and are being implemented in many fields. They can be broadly divided into unsupervised learning and supervised learning. Unsupervised learning methods aim to extract hidden structures within the neural data, commonly used for feature extraction, pattern recognition, clustering, and dimensionality reduction. Supervising learning methods train the neural data using underlying functions to map to a given output and automatically discover the relationships between input data and output labels. The most common applications of supervised learning are classification and regression.

Quantitative models of machine learning algorithms provide incredibly powerful implementations in neuroscience. Some traditional methods such as LDA, PCA, and support vector machine (SVM) are also regarded as machine learning algorithms. Other algorithms (neural networks, autoencoders, and logistic regression) train batches of input data using basis transformation function to match the output adaptively.

4. Applications of neural signal processing

Neuroprostheses, neurostimulators, or human-machine interfaces are devices that record from or stimulate the brain to help individuals with neurological disorders, restore their lost function, and thereby improve their quality of life. Neural signal processing methodologies are used extensively in all these applications.

4.1 Neurostimulators

Neurostimulators that have demonstrated decades of success are cochlear implants that are designed for those who have dysfunctional conduction of sound waves from the eardrum to the cochlea. These implants can also help elderly individuals who have age-related hearing loss [4]. There is an external speech processor to capture and convert the sound from the surrounded environment to digital signals. The internal implants turn the digital signals into electrical signals to stimulate the hearing nerve by the electrodes inside the cochlea. Once the brain receives the signals, one can hear and interpret the sound.

Another successful neurostimulator is the deep brain stimulator (DBS) system used for individuals with Parkinson's disease. The DBS has been available as a reliable treatment for decades for individuals with Parkinson's disease. The implanted impulse generator placed under the collarbone provides continuous electrical impulses by giving a certain frequency of stimulation to the subthalamic nucleus and makes it possible to minimize the uncontrolled tremors. During the DBS surgery, electrodes are inserted into a targeted area of the brain, and the whole procedure is monitored and recorded using MRI. After the treatment, symptomatic improvement was durable for at least 10 years [5].

4.2 Neuroprostheses or human-machine interfaces (HMIs)

Stroke, spinal cord injury, and traumatic brain injury may lead to long-term disability, and an increased number of individuals are suffering from severe motor impairments, resulting in loss of independence in their daily life. Recovery of motor function is crucial in order to perform activities of daily living. Human-machine interfaces (HMIs) can enable dexterous control of exoskeletons that could be used as a rehabilitative device or an assistive device to restore lost motor functions poststroke or spinal cord lesions [6–8], thus promoting long-lasting improvements in motor function of individuals with movement disorders.

Additionally, significant applications in neural engineering are HMI-based systems to restore or compensate the lost limb functions for individuals with amputation or paralysis. Cortical control of prosthetics has been studied both in animals [9] and humans [10, 11]. Movement-related cortical potentials used to assess cortical activation patterns provide interesting information, as they are associated with the planning and execution of voluntary movements. Recently, HMI-based research has stressed on the development of algorithms for movement decoding using noninvasive neural recordings [12–14]. In order to understand neural intent before or during the movement, it is necessary to extract the characteristics accurately using efficient algorithms. The adaptability and reliability over the long-term are current challenges that are being addressed using advanced and adaptive signal processing methodologies. Months to years of training are essential to operate prosthetic or exoskeleton skillfully. This training time can possibly be reduced by increasing the burden on machine learning algorithms that are currently being addressed by advanced signal processing methods.

4.3 Neurological disorders

Epilepsy is a common neurological disorder characterized by an enduring predisposition to generate epileptic seizures [15]. These seizures may cause disturbances in movement, loss of control of bowel or bladder function, loss of consciousness, or other disturbances in cognitive functions. Currently, the signal processing algorithms can detect ongoing seizures and provide clinicians with detailed information such as localization of seizure foci useful for the treatment of epilepsy. The ability to detect seizures rapidly and accurately could promote therapies aimed at rapid treatment of seizures.

Skilled neurophysiologists visually examine the neural signals and detect epilepsy. Apart from the single-channel signals, other contextual information such as spatial and temporal data are vital to neurophysiologists for recognizing spikes [16, 17]. Currently, the epileptic seizures can be detected and predicted from EEG or ECoG signals by extracting the hidden features using machine learning algorithms [15, 18–20].

5. Conclusion

Neural signal processing has become an increasingly important tool in neuroscience and neural engineering. This chapter provides a general overview of the widely implemented neural recording and stimulation technologies, neural processing methodologies, and how these techniques can be practically executed in some practical applications. The understanding of the neural representations of human behavior in the brain can be significantly enhanced using neural signal processing methodologies. With the advances in neural recording techniques and parallel advances in neural signal processing, we believe that several unanswered questions and challenges in neuroscience and neural engineering will be solved in the near future.

Author details

Dingyi Pei and Ramana Vinjamuri*
Department of Biomedical Engineering, Stevens Institute of Technology, United States

*Address all correspondence to: ramana.vinjamuri@gmail.com

References

[1] Wonham W. The internal model principle of control theory. Automatica. 1976;**12**(5):457-465

[2] Kawato M, Wolpert D. Internal models for motor control. In: Sensory Guidance of Movement; 1998. pp. 291-307

[3] Fransen AMM, van Ede F, Maris E. Identifying neuronal oscillations using rhythmicity. NeuroImage. 2015;**118**: 256-267

[4] Cabral Jr F, Pinna MH, Alves RD, Malerbi AFDS, Bento RF. Cochlear implantation and single-sided deafness: A systematic review of the literature. International Archives of Otorhinolaryngology. 2014;**20**(1):69-75

[5] Hitti FL, Ramayya AG, Mcshane BJ, Yang AI, Vaughan KA, Baltuch GH. Long-term outcomes following deep brain stimulation for Parkinson's disease. Journal of Neurosurgery. 2019;**132**(1):1-6

[6] Monge-Pereira E, Ibañez-Pereda J, Alguacil-Diego IM, Serrano JI, Spottorno-Rubio MP, Molina-Rueda F. Use of electroencephalography brain-computer interface systems as a rehabilitative approach for upper limb function after a stroke: A systematic review. PM & R: The Journal of Injury, Function, and Rehabilitation. 2017;**9**(9):918-932

[7] Frolov AA, Mokienko O, Lyukmanov R, Biryukova E, Kotov S, Turbina L, et al. Post-stroke rehabilitation training with a motor-imagery-based brain-computer interface (BCI)-controlled hand exoskeleton: A randomized controlled multicenter trial. Frontiers in Neuroscience. 2017;**11**(JUL)

[8] Burns MK, Pei D, Vinjamuri R. Myoelectric control of a soft hand exoskeleton using kinematic synergies. IEEE Transactions on Biomedical Circuits and Systems. 2019;**13**(6):1351-1361

[9] Velliste M, Perel S, Spalding MC, Whitford AS, Schwartz AB. Cortical control of a prosthetic arm for self-feeding. Nature. 2008;**453**(7198):1098-1101

[10] Yanagisawa TT, Hirata M, Saitoh Y, Kishima H, Matsushita K, Goto T, et al. Electrocorticographic control of a prosthetic arm in paralyzed patients. Annals of Neurology. 2012;**71**(3):353-361

[11] Bright D, Nair A, Salvekar D, Bhisikar S. EEG-based brain controlled prosthetic arm. Conference on Advances in Signal Processing (CASP) 2016. 2016. pp. 479-483

[12] Pei D, Burns M, Chandramouli R, Vinjamuri R. Decoding asynchronous reaching in electroencephalography using stacked autoencoders. IEEE Access. 2018;**6**:52889-52898

[13] Pei D, Patel V, Burns M, Chandramouli R, Vinjamuri R. Neural decoding of synergy-based hand movements using electroencephalography. IEEE Access. 2019;**7**:18155-18163

[14] Patel V, Burns M, Pei D, Vinjamuri R. Decoding synergy-based hand movements using electroencephalography. In: 40th Annual International Conference of the IEEE Engineering in Medicine and Biology Society (EMBC). 2018. pp. 4816-4819

[15] Elger CE, Hoppe C. Diagnostic challenges in epilepsy: Seizure under-reporting and seizure detection. Lancet Neurology. 2018;**17**(3):279-288

[16] James CJ, Jones RD, Bones PJ, Carroll GJ. Detection of epileptiform

discharges in the EEG by a hybrid system comprising mimetic, self-organized artificial neural network, and fuzzy logic stages. Clinical Neurophysiology. 1999;**110**(12):2049-2063

[17] Tzallas AT, Oikonomou VP, Fotiadis DI. Epileptic spike detection using a Kalman filter based approach. Conference Proceedings: Annual International Conference of the IEEE Engineering in Medicine and Biology Society. 2006. pp. 501-504

[18] Ramgopal S, Thome-Souza S, Jackson M, Kadish NE, Fernández IS, Klehm J, et al. Seizure detection, seizure prediction, and closed-loop warning systems in epilepsy. Epilepsy & Behavior. 2014;**37**:291-307

[19] Naghsh-Nilchi AR, Aghashahi M. Epilepsy seizure detection using eigen-system spectral estimation and multiple layer perceptron neural network. Biomedical Signal Processing and Control. 2010;**5**(2):147-157

[20] Chakrabarti S, Swetapadma A, Pattnaik PK. A Review on Epileptic Seizure Detection and Prediction Using Soft Computing Techniques. Vol. 374. Springer International Publishing; 2019

Chapter 2

Cerebral Spectral Perturbation during Upper Limb Diagonal Movements

Fabio Marson, Patrizio Paoletti, Stefano Lasaponara, Joseph Glicksohn, Antonio De Fano and Tal Dotan Ben-Soussan

Abstract

While it has been suggested that diagonal rhythmical bilateral movements promote improvement in motor and cognitive functions, no study that we are aware of has actually examined electrophysiological changes during diagonal movements. Therefore, we aimed to study cerebral activity during the performance of diagonal and vertical movements (DM and VM, respectively), through EEG recording focusing on theta, alpha, and beta frequency bands. Following independent component analysis, we computed time-frequency and source localization analysis. We found that (1) increased frontal theta during the initiation of DM was possibly related to the computational effort; (2) a biphasic pattern of frontoparietal alpha/beta modulations was found during VM; and in addition, (3) source localization showed increased frontal theta during DM generated in the middle frontal cortex. We will discuss the current results and their implications in relation to task difficulty, spatial and temporal computation.

Keywords: diagonal movement, ERSP, source localization, time-frequency analysis

1. Introduction

1.1 Diagonal movement overview

Diagonal arm movements are upper limb movements performed along the sagittal plane that implies the vertical body midline in order to reach the opposite hemispace. For this reason, diagonal movements are inherently different from movements performed along either the transverse or vertical planes. In fact, during vertical and horizontal movements, upper limbs cross the transverse and vertical body midline, respectively, but not both of these axes at the same time. In fact, Adele Diamond [1] has suggested the idea that rhythmical, diagonal, bilateral coordinated arm movements and eye-hand coordination could play a role in cognitive rehabilitation as well, providing a framework that could reduce the distance between the idea that motor activity cannot influence cognition and that cognitive rehabilitation requires only pure cognitive-related protocols.

At the beginning of the twenty-first century, Diamond summarized different scientific evidence to highlight the fact that movement and cognition are linked in terms of development timing and that these two domains activate shared brain regions such as the prefrontal cortex and cerebellum [2]. Specifically, it was observed that activity in the prefrontal cortex increases when a motor task requires the maintenance of movement information through execution, selection of relevant task-related motor information, and inhibition of automated behavior that could compromise motor performance.

Yet very few studies have examined this issue. A rare exception regards the studies related to a rehabilitation protocol based on stretching exercise and joint mobility called proprioceptive neuromuscular facilitation (PNF) [3]. This rehabilitation protocol also includes diagonal arm movements in order to promote strength, coordination, functional motoricity, and movement initiation [4]. More recently, Moreira et al. [5] studied cerebral electrophysiological activity after the performance of unimanual diagonal arm movements as presented in the original PNF protocol. They observed an increase of beta and alpha power within a frontoparietal network after performance. They showed that this kind of motor activity could improve motoricity through modulation of cerebral plasticity over frontoparietal regions, suggesting that improved motor execution is directly linked to motor control in the central nervous system.

Over the last few years, we studied another motor task that involves diagonal movements, named Quadrato Motor Training (QMT) [6–8]. QMT is a sensorimotor training based on step-by-step, whole-body movement in vertical, horizontal, and diagonal directions. From the electrophysiological point of view, pre- and post-EEG recording showed acute and chronic increase of theta and alpha power and coherence over frontal, parietal, and temporal regions [9–11]. We hypothesized that it is especially the planning of diagonal movement which plays a crucial role in the establishment of such a modulation of theta and alpha band activity. However, it is difficult to disentangle and isolate the contribution of diagonal from vertical and horizontal movement, and further studies in which EEG is recorded during the execution of QMT should be performed. Thus, as a first step, in order to examine the effects of diagonal movements, in the current study, we measured cerebral activity during the performance of diagonal and vertical movements (DM and VM, respectively), using EEG recording. Given the aforementioned literature, particular focus was addressed to the theta, alpha, and beta frequency bands.

1.2 Brain oscillations, movement, and cognition

Theta (4–7 Hz) activity seems to be involved in different cognitive functions such as sustained attention [12], spatial navigation [13], memory [14], meditative states/internalized attention [15], and creativity [16]. Importantly, all these cognitive functions require integration from different cerebral regions in order to produce effective outcomes. In fact, it was suggested that there is an inverse relationship between the extent of a recruited cortical network and the elicited oscillatory frequency during task performance [17]. For these reasons, theta is also thought to support long-range integration and promotion of mental states related to absorption and concentration [18–20].

Alpha activity (8–12 Hz) is considered a fundamental brain rhythm produced in the occipital cortex which reflects cortical inactivity during relaxed wakefulness with closed eyes and reduced sensory and motor processing [21, 22]. Alpha desynchronization has been observed during a task that requires the deployment of attention toward specific targets or locations in space, suggesting that alpha could play an important role in the management of attentional resources and sensory perception [23–26]. Moreover, similar to theta, increased parietal alpha power has been related to internal-directed attention [27].

Beta frequency (13–30 Hz) is classically related to active wakefulness. It was observed as an oscillatory activity replacing alpha waves when individuals opened their eyes [28–30]. Beta desynchronization has been typically observed during the execution of voluntary movement [31, 32] and during sensorimotor processing [33]. After the conclusion of a voluntary movement, beta synchronization called "post-movement beta rebound" has been observed over the sensorimotor cortex [34, 35]. Other studies have suggested that beta is also involved in a large range of cognitive, emotional, and attentional processing [36–39].

1.3 Aim of this study

Despite the fact that movement along the diagonal axis has been seen as a benefit in different motor and cognitive rehabilitation protocols, none of the studies we are aware of have investigated the neural correlates of diagonal movement during the actual movement performance. Therefore, we aimed to observe the electrophysiological correlates of diagonal movements and compare them to a control condition such as vertical movement. We hypothesized that diagonal movements compared to vertical movements require more computational effort and motor control to be well executed. This should result in increased theta, in decreased alpha due to an increase in internalized attention, and finally in a modulation of beta tied to the start and the end of each movement.

2. Methods

2.1 Participants and design

Eleven right-handed healthy participants (six males; mean age 40.3 years; SD 9.22) volunteered to take part in the study. All participants had normal or corrected-to-normal vision, and they declared to not suffer from any psychiatric or physical disease that could interfere with their performance.

The research took place in the Research Institute for Neuroscience, Education, and Didactics of the Patrizio Paoletti Foundation. The participants signed an informed consent. The study was approved by the ethics committee of Bar-Ilan University.

Before each experimental session, participants performed a training phase in which they learned to perform correctly the movements required for the experiment. During the experimental session, EEG recording was conducted during the performance of diagonal and vertical movements with both arms at the same time in an antiphase movement (i.e., each arm moved jointly with the other arm starting from the opposite side of the body and moving toward the opposite direction along the same axis; for a clear explanation, see **Figure 1**).

2.2 Paradigm

Participants performed a total of 320 rhythmical movements with both arms while sitting in a chair. Movements were divided into eight blocks. Each block consisted in 40 rhythmical continuous movements paced by an external sound. We used a 440 Hz tone with a duration of 100 ms, presented with the pace of 1 Hz (i.e., one sound each second), in order to guide the movements.

At the beginning of each block, participants had to keep one arm in the upper position and the other one in the lower position (see **Figure 1**). Then, in line with the start of the pacing sounds, they had to move their arms along the vertical or diagonal axis, according to the condition of the block. A total of 20 consecutive trials were used in each block. Each trial lasted 2 s. Therefore, each trial was composed of two sounds (i.e., two movements). The participants were instructed to perform

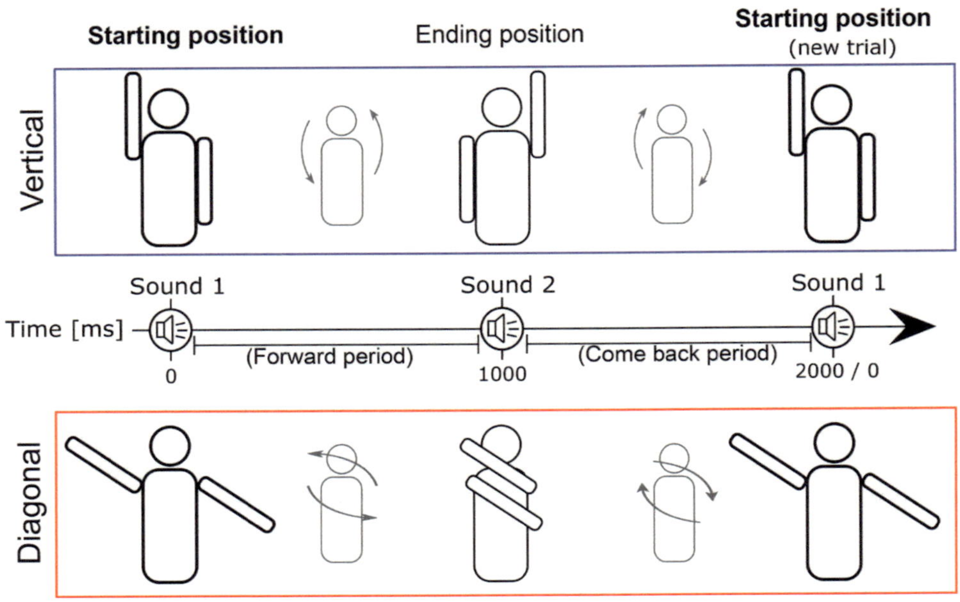

Figure 1.
Examples of the structure of a single trial. Each trial lasts 2 s and is composed of two sounds played at 0 and 1000 ms. Since each block was composed of 20 continuous movements, the sound at 2000 ms of the represented trial is indeed the sound at 0 ms of the next trial. Here vertical and diagonal conditions are both depicted with the right arm in the upper position.

the movement between each sound in order to be in the starting position or in the ending position whenever a sound occurred.

Four experimental conditions (movement (diagonal, vertical) and hand (right, left)) were used: (1) vertical movement starting with the right hand above head, (2) diagonal movement starting with the right hand above head, (3) vertical movement starting with the left hand above head, (4) and diagonal movement starting with the left hand above head. Each condition was repeated twice.

2.3 EEG

2.3.1 EEG recording

EEG was recorded using the eego sports system from ANT Neuro (Enschede, Netherlands). We recorded electrophysiological brain activity from 32 scalp sites with a 10/10 electrode layout. In addition, we used a three-dimensional accelerometer placed on the right arm in order to record movements along x, y, and z axes. Recording was conducted with a sample rate of 1000 Hz, and impedance was kept below 10 kohm. Online reference was CPz.

The experiment was conducted in a dimly lighted room and implemented in E-Prime 2.0 (Psychology Software Tools, Pittsburgh, PA). We used a standard computer monitor for visual output and two speakers placed below the screen for sound presentation.

2.3.2 EEG analysis

EEG signals were preprocessed and analyzed using EEGLAB toolbox [40] for time-frequency analysis and eLORETA for source localization analysis. First, data were preprocessed using a 50 Hz notch filter and filtered between 1 and 40 Hz with a second-order infinite impulse response (IIR) Butterworth filter. The first two trials from each block were rejected in order to avoid transient activity related to the start of the continuous movement.

We then performed independent component analysis (ICA) using the Infomax algorithm implemented in EEGLAB, over the whole set of electrodes along the whole recording. Through ICA, we deconstructed the signal into 34 independent components, allowing us to identify and reject ocular and major motion artifacts. After labeling and rejection of non-brain-derived activity, we projected the components back into the channel domain to obtain clear EEG time course and to perform further analysis (for ICA methodological information, see [41]). EEG was finally offline referenced against the mean of all derivations.

2.3.2.1 Time-frequency analysis

Two-second time windows were extracted, locking t0 with the first sound of each trial.

Time-frequency analysis was conducted for both single electrodes and four different region of interests (ROIs): frontal (electrodes F3, Fz, F4), fronto-central (electrodes FC1, FC2), parietal (electrodes P3, Pz, P4), and POz. These ROIs were selected on the base of a previous experiment [5] in which an increase in alpha and beta activity was found along the frontoparietal axis after the performance of diagonal movements. We extracted time-frequency courses for theta (4–7 Hz), alpha (8–12 Hz), and beta (13–30 Hz) frequency bands.

First, we computed event-related spectral perturbation (ESRP) on the whole trial period in EEGLAB. In this way, we obtained one matrix for each electrode, hence a total of 32 matrices. Each 100 by 200 matrix was composed of 100 frequency values (from 1 to 50 Hz with frequency values distributed logarithmically over the total amount of rows) × 200 time points (from 1 to 2000 ms). A similar analysis was conducted by Cohen and colleagues [42].

Then, in order to analyze power value change in each frequency band of interest, we extracted and averaged data for each desired frequency band, comparing vertical and diagonal arm movement-related spectral perturbation over time.

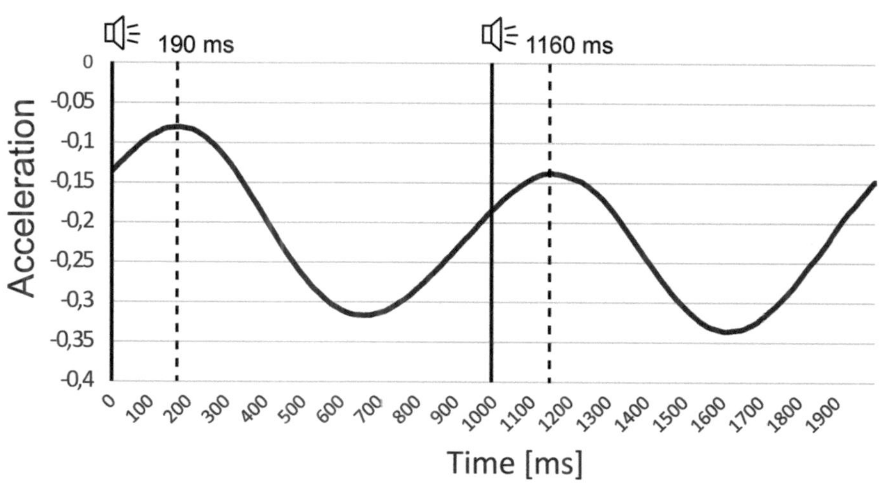

Figure 2.
Accelerometer data representing peaks of movement as a function of time (resulting from the movement). We averaged data from all trials in vertical and diagonal conditions in order to identify the time points in which the arm reached the starting and the ending point of each movement. Bold lines represent presentation of the pacing sounds.

The trial period was split into two time windows: forward period (1000 ms after the first sound (1–1000 ms)) and comeback period (1000 ms after the second sound for each trial (1001–2000 ms)). These two time windows represent two distinct movements in each trial. Forward period includes the movement from the starting position to the ending position, while comeback period includes the movement back from the ending position to the starting position (see **Figure 1**).

Statistical analysis was performed on specific time windows in both forward period and comeback period. Time windows were first defined by the accelerometer data. Our accelerometer can detect not only the acceleration but also the position in space starting from a baseline reference position. We observed that the participants reached the starting point and ending point slightly after the actual sound presentation. We extracted data from the accelerometer from each trial and observed that the delay between sound presentation and actual start of the movement from the starting point was 190 ms while between sound presentation and start of actual movement from the ending point was 160 ms (see **Figure 2**).

Finally, statistical analysis was performed using STATISTICA software (StatSoft, Inc., Tulsa, OK, USA).

2.3.2.2 Source localization

Using low-resolution brain electromagnetic tomography (LORETA) [43], it is possible to solve the inverse problem in EEG and localize generators of electrophysiological components of EEG signal in a specific frequency band.

After time-frequency analysis, we performed source localization analysis using LORETA in order to observe differences between vertical and diagonal movements for theta (4–7 Hz), alpha (8–12 Hz), and beta (13–30 Hz) activity. Source localization analysis was conducted in specific time windows using a data-driven approach, according to what we observed in the time-frequency analysis.

More specifically, we focused on the time windows previously observed in time-frequency analysis for each specific frequency band. Therefore, we compared the generator of theta in diagonal and vertical movements during the planning of movement in forward period (between 100 and 300 ms), of alpha during the two peaks of activity in comeback period (200–400 ms; 650–850 ms) and of beta in forward period (320–520 ms) and comeback period (220–420 ms). Specifically, analyzed frequencies were theta (7 Hz), alpha (11 Hz), and beta (two frequencies, 19 and 23 Hz).

We performed one-tailed t-test comparisons based on the time-frequency observed activity pattern (i.e., diagonal-related activity greater than vertical-related activity or vice versa). Therefore, for theta, we expected diagonal > vertical; for alpha, we expected diagonal > vertical in P1 and vertical > diagonal in P2; and for beta, we expected diagonal > vertical in both time windows.

Statistical analysis was conducted using subject-wise normalization, and results are expressed as a t-test on the logarithmically transformed data. Nonparametric randomized permutation and probability threshold corrections were performed [44].

3. Results

3.1 Time-frequency results

3.1.1 Theta (4–7 Hz)

We selected a time window ranging from −100 to +100 ms around the peaks recorded by accelerometer and then extracted and analyzed the activity. We identified the two peaks in forward period and comeback period as peak forward (190 ms) and peak comeback (1150 ms), respectively.

We performed a 2 × 2 ANOVA with condition (vertical, diagonal) and peak (peak forward, peak comeback) as factors. ANOVA was performed for specific ROIs: frontal ROI (F3, Fz, F4), Fz, and parietal ROI (P3, Pz, P4). We performed analysis on Fz separately because we know that this region is particularly relevant for the generation of frontal midline theta [12, 15].

For the frontal ROI, none of the main effects was significant (all $p > 0.14$). The condition-peak interaction was significant [$F(1, 10) = 5.21$; $p < 0.05$; $\eta 2 = 0.34$]. Fisher LSD post hoc analysis showed that peak forward in the diagonal condition was significantly higher than peak forward in the vertical condition ($p < 0.03$) and from peak comeback in the diagonal condition ($p < 0.05$) (see **Figure 3**).

For channel Fz, none of the main effects was significant ($p > 0.11$). The condition-peak interaction was significant [$F(1, 10) = 5.15$; $p < 0.05$; $\eta 2 = 0.34$]. Fisher LSD post hoc analysis showed that peak forward in the diagonal condition was significantly different from peak forward in the vertical condition ($p < 0.03$) and from peak comeback in the diagonal condition ($p < 0.03$) (see **Figure 3**).

For the parietal ROI, none of the main effects or their interaction was significant (all $p > 0.14$).

3.1.2 Alpha (8–12 Hz)

First, we analyzed the same time window previously analyzed for the theta band (Peak1 = 90–290 ms, Peak2 = 1060–1260 ms) in the same clustered ROIs (frontal ROI, parietal ROI) and POz (see **Figure 4**).

For all ROIs, no main effects or their interaction was significant (all $p > 0.15$).

In order to conduct a more specific analysis, we observed the time course and performed qualitative assessment by visual inspection. In this way, we were able to divide each period into time windows to be used for analysis. We used this manual selection method to identify specific time windows of interest in a data-driven approach

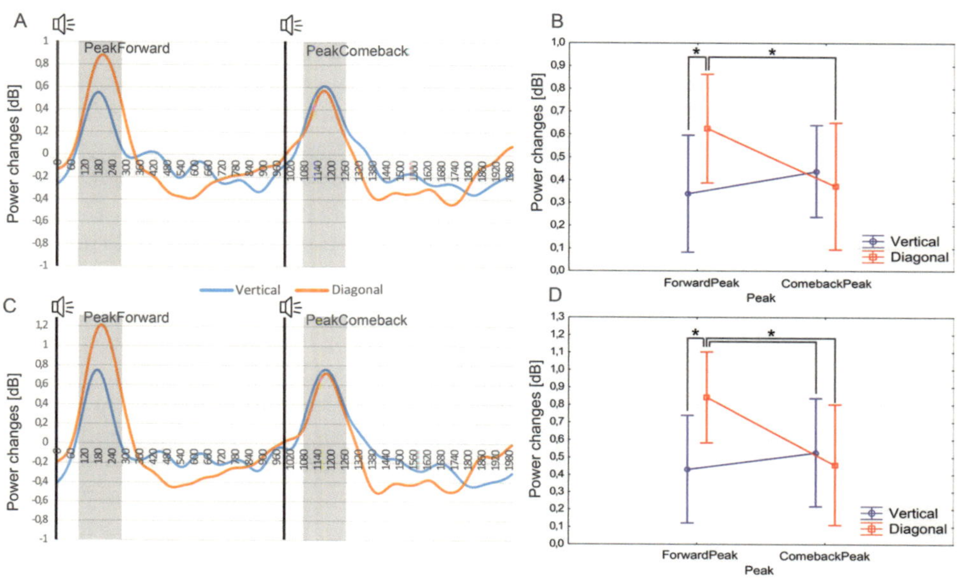

Figure 3.
Theta (4–7 Hz) time-frequency plot (left panels) and analysis (right panels) of the 2 × 2 ANOVA with condition and peak as factors in frontal ROI (A, B) and Fz (C, D). In A and C, highlighted rectangles represent time windows used for analysis. In B and D, bold lines represent significant differences (p < 0.05).

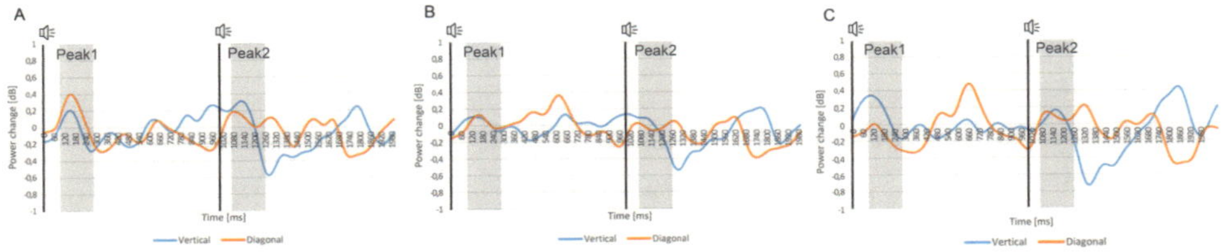

Figure 4.
Alpha (8–12 Hz) time-frequency plot with time windows of interest for frontal ROI (A), parietal ROI (B), and POz (C).

(a similar approach was used by Missonnier et al. see [45]). Maclin et al. [46] used a similar approach to manually select specific noncanonical frequency bands of interest. After visual inspection, we decided to adopt different approaches for each period, basing our criterion on specific parameters that will be explained later.

3.1.2.1 Alpha: forward period

After visual inspection, since we did not have an a priori hypothesis, we decided to analyze forward period adopting an exploratory approach. We divided the forward period into four time windows of 250 ms each in order to analyze the computation of the movement (0–250 ms; t1), the start of the movement (250–500 ms; t2), the crossing of body midline (500–750 ms; t3), and the final part of the movement execution (750–1000 ms; t4) (see **Figure 5A**, **C** and **E**).

We employed a 2 × 4 ANOVA using condition (vertical, diagonal) and time window (t1, t2, t3, t4) as factors. We performed analysis on frontal and parietal ROIs.

None of the main effects or their interaction were significant for frontal ROI (all $p > 0.09$), for parietal ROI (all $p > 0.18$), and for POz (all $p > 0.13$).

3.1.2.2 Alpha: comeback period

Observing the time course during comeback period in the alpha range, we noticed two distinct peaks of activity both in frontal and parietal regions, located at 1290 and 1790 ms, respectively, after the start of the comeback period.

First, we segmented the comeback period into three main stages in order to analyze the two main peaks and the activity between them. The external time windows last 200 ms and fell around the two main peaks. Between these two time windows, we identified a central additional one. Therefore, we identified the three time windows as follows: p1 = 200–400 ms, p2 = 400–700 ms, p3 = 700–900 ms (see **Figure 5B**, **D** and **F**).

We performed a 2 × 3 ANOVA using condition (vertical, diagonal) and time window (p1, p2, p3) as factors. We performed the analysis on frontal and parietal ROIs and POz.

For the frontal ROI, none of the main effects were significant (all $p > 0.30$). The condition-time window interaction was significant [$F(2, 20) = 4.37$; $p < 0.05$; $\eta 2 = 0.30$]. Fisher LSD post hoc revealed that p1 in vertical was significantly different from p3 in vertical ($p < 0.05$), while all other comparisons were not significant (see **Figure 6A**).

For the parietal ROI, none of the main effects were significant (all $p > 0.59$). The condition-time window interaction was significant [$F(2,20) = 6.13$; $p < 0.01$; $\eta 2 = 0.38$]. Fisher LSD post hoc revealed that p1 in vertical was significantly different from p3 in vertical ($p < 0.05$), p1 in diagonal ($p < 0.05$), and p2 in diagonal ($p < 0.05$), while all other comparisons were not significant. Comparison between vertical and diagonal condition was significant in p1 ($p < 0.05$) and p3 ($p < 0.05$) (see **Figure 6B**).

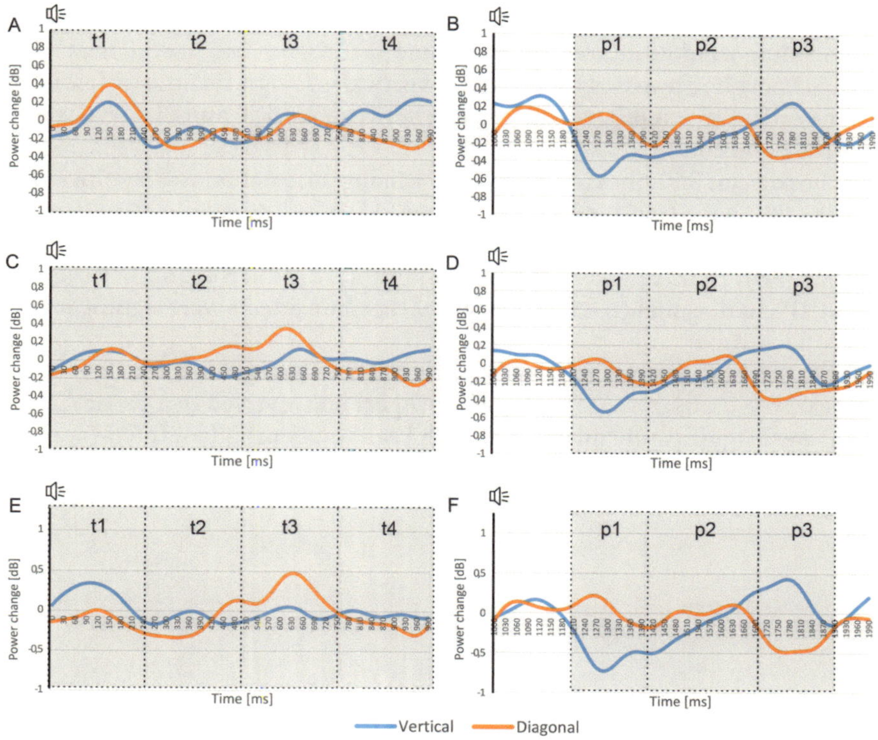

Figure 5.
Alpha (8–12 Hz) time-frequency plot with time window segmentation of forward period (left side) in frontal ROI (A), parietal ROI (C), and POz (E) and comeback period (right side) in frontal ROI (B), parietal ROI (D), and POz (F).

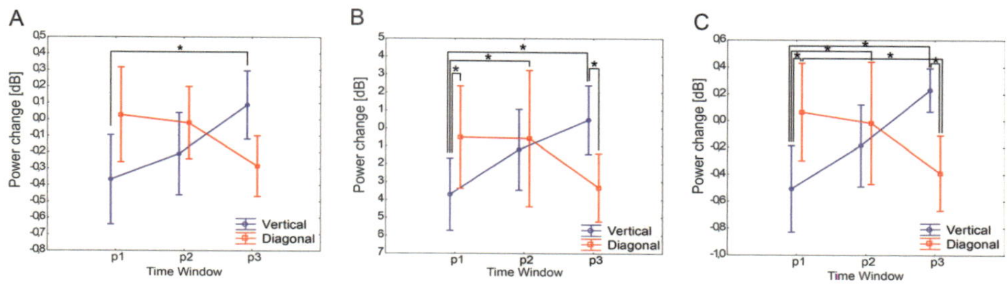

Figure 6.
Alpha (8–12 Hz) analysis of time windows p1, p2, and p3 during comeback period in frontal ROI (a), parietal ROI (B), and POz (panel C).

For POz, none of the main effects were significant (all $p > 0.57$). The condition-time window interaction was significant [$F(2, 20) = 8.64$; $p < 0.01$; $\eta 2 = 0.46$]. Fisher LSD post hoc revealed that p1 in vertical was significantly different from p3 in vertical ($p < 0.01$), p1 in diagonal ($p < 0.05$), and p2 in diagonal ($p < 0.05$); p3 in diagonal was significantly different from p3 in vertical ($p < 0.01$) and p1 in diagonal ($p < 0.05$) (see **Figure 6C**).

3.1.2.3 Alpha: single peaks

To refine our analysis, we extracted the activity in the two main peaks using a time window of 50 ms around the maximum value (Peak1 = 240–340 ms; Peak2 = 740–840 ms) and performed a 2 by 2 ANOVA with condition (vertical, diagonal) and peaks (peak forward, peak comeback) as factors (see **Figure 7**).

For the frontal ROI, none of the main effects were significant (all $p > 0.32$). The condition-time window interaction was found [$F(2,20) = 7.66$; $p < 0.05$; $\eta 2 = 0.43$]. Fisher LSD post hoc revealed that Peak1 in vertical was significantly different from Peak2 in vertical ($p < 0.05$). No differences between vertical and diagonal were found (see **Figure 7A**).

For the parietal ROI, the condition-time window interaction was significant [$F(2, 20) = 28.72$; $p < 0.001$; $\eta 2 = 0.74$]. Fisher LSD post hoc revealed that vertical and diagonal were significantly different in both Peak1 and Peak2 ($p < 0.01$) (see **Figure 7B**). Peak1 and Peak2 were also significantly different for both vertical ($p < 0.001$) and diagonal ($p < 0.05$). None of the main effects were significant (all $p > 0.07$).

For POz, the condition-time window interaction was significant [$F(2, 20) = 28.52$; $p < 0.001$; $\eta 2 = 0.74$]. Fisher LSD post hoc revealed that vertical and diagonal were significantly different in both Peak1 and Peak2 ($p < 0.01$) (see **Figure 7C**). Peak1 and Peak2 were significantly different in both vertical ($p < 0.001$) and diagonal ($p < 0.05$). None of the main effects were significant (all $p > 0.19$).

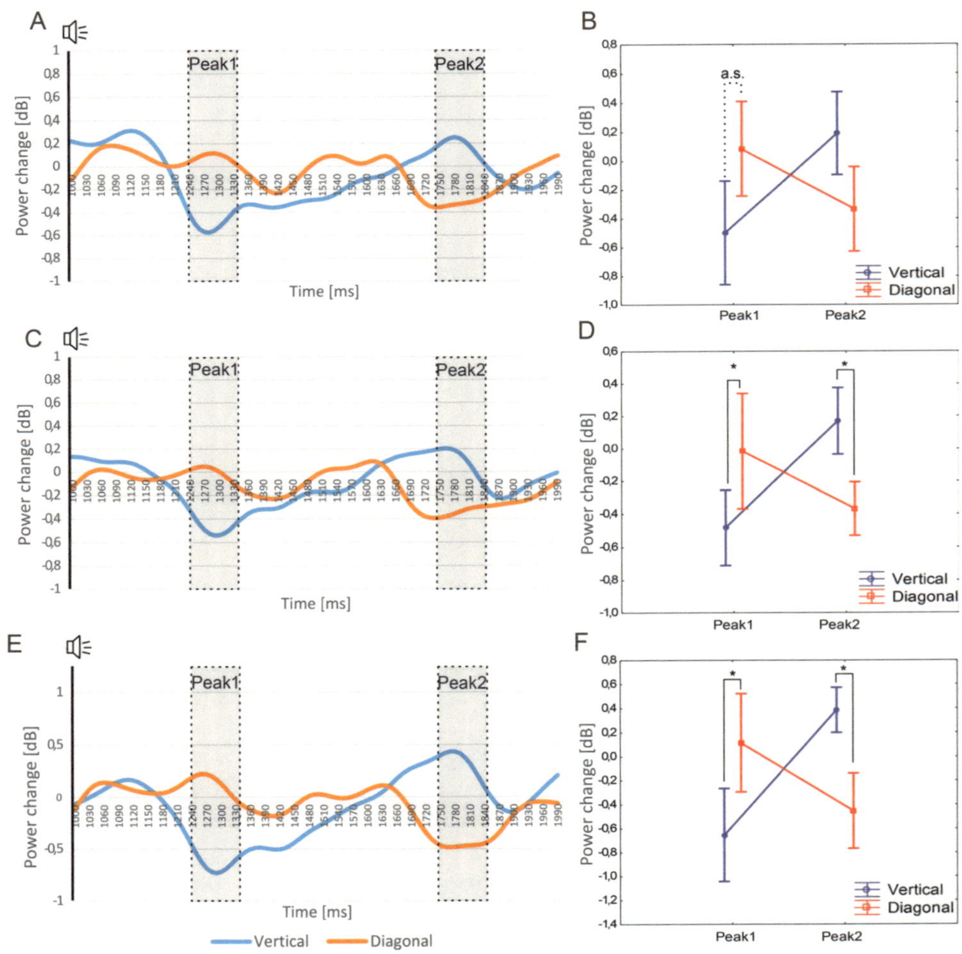

Figure 7.
Alpha (8–12 Hz) time-frequency plot with peak segmentation during comeback period in frontal ROI (A), parietal ROI (C), and POz (E). Analysis of peaks p1 and p2 during comeback period in frontal ROI (B), parietal ROI (panel D), and POz (panel F). In A, C, and E, highlighted rectangles represent time windows used for analysis. In B, D, and F, only significant or almost significant comparisons between vertical and diagonal are displayed. Bold lines represent significant differences (p < 0.05), and the dashed line represents a difference that approaches statistical significance.

3.1.2.4 Alpha1 vs. Alpha2

Since lower and upper alphas have previously been reported to have different roles, lower alpha being related to general attentional processes and upper alpha being more sensitive to perceptual and semantic processing during task performance [51], we further wanted to explore the possibility that lower and upper alpha could have different involvement in the modulation of alpha biphasic activity during vertical movements. Therefore, we analyzed POz, where we have found the maximum alpha modulation.

We divided the alpha band into Alpha1 and Alpha2 (8–10 Hz and 11–12 Hz, respectively) and performed a two-way ANOVA with alpha (Alpha1, Alpha2) and peak (Peak1, Peak2) as factors.

The main effect of peak was significant [$F(1, 10) = 11.42$; $p < 0.01$; $\eta 2 = 0.53$], suggesting that Peak2 (mean, 0.432 dB) was significantly higher than Peak1 (mean. −0.689 dB). The alpha-peak interaction was significant [$F(1, 10) = 14.5$; $p < 0.01$; $\eta 2 = 0.59$]. Fisher LSD post hoc revealed that comparison between Alpha1 and Alpha2 approached significance in Peak1 ($p = 0.07$) and was significant in Peak2 ($p < 0.05$) (see **Figure 8**).

3.1.3 Beta (13–30 Hz)

For analysis in the beta frequency, we used the same time window already used for the alpha band. In forward period (M1), we therefore used four consecutive time windows lasting 250 ms each, while in the comeback period, we used three time windows: p1 = 200–400 ms, p2 = 400–700 ms, p3 = 700–900 ms (see **Figure 9**).

3.1.3.1 Beta: forward period

A 2 × 4 ANOVA using condition (vertical, diagonal) and time window (t1, t2, t3, t4) as factors was conducted. We performed analysis on the frontal and parietal ROIs. For the frontal ROI, none of the main effects or interaction were significant (all $p > 0.25$). For the parietal ROI, none of the main effects were significant (all $p > 0.38$).

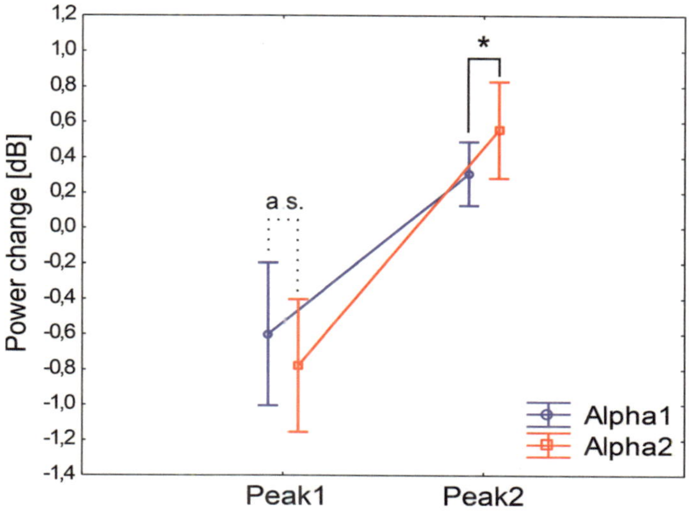

Figure 8.
Comparisons between Alpha1 and Alpha2 in Peak1 and Peak2. The bold line represents a significant difference (p < 0.05), and the dashed line represents a difference that approaches statistical significance.

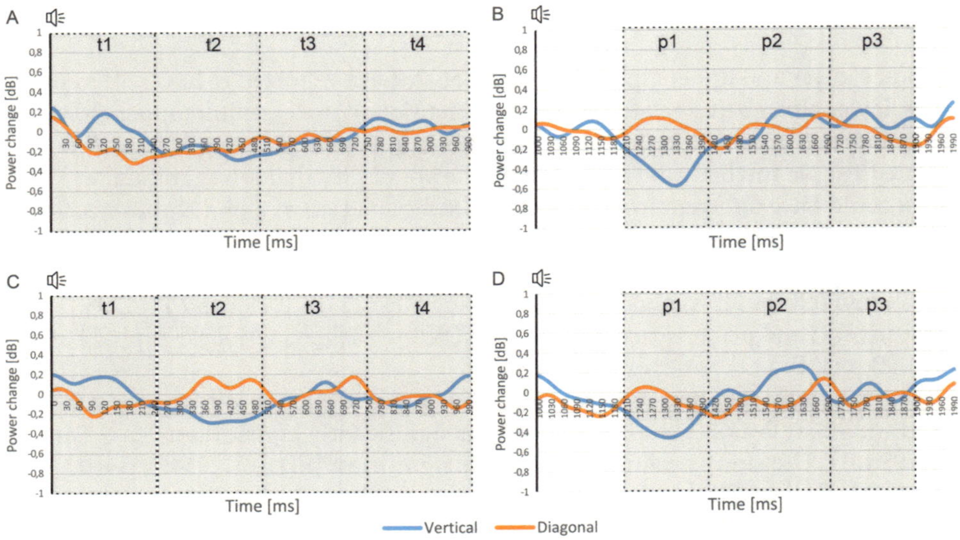

Figure 9.
Beta (13–30 Hz) time-frequency plot with peak segmentation during forward period (left panels) and comeback period (right panels) for frontal ROI (A, B) and parietal ROI (C, D).

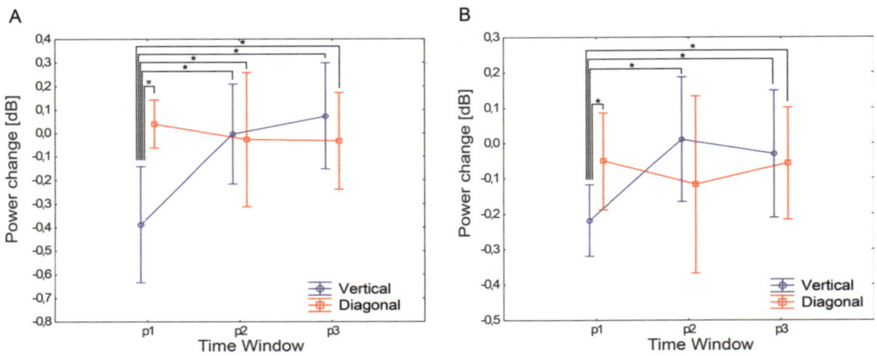

Figure 10.
Beta (13–30 Hz) analysis of time windows p1, p2, and p3 during comeback period for frontal ROI (A) and parietal ROI (B). Bold lines represent significant differences ($p < 0.05$).

3.1.3.2 Beta: comeback period

We performed a 2 × 3 ANOVA using condition (vertical, diagonal) and time window (p1, p2, p3) as factors. We performed the analysis on frontal and parietal ROIs.

For the frontal ROI, none of the main effects were significant (all $p > 0.12$). The condition-time window interaction was significant [$F(2, 20) = 4.07$; $p < 0.05$; $\eta 2 = 0.28$]. Fisher LSD post hoc revealed that p1 in vertical was significantly different from all other factors ($p < 0.05$) (see **Figure 10A**).

For the parietal ROI, none of the main effects were significant (all $p > 0.55$). The condition-time window interaction was significant [$F(2, 20) = 5.45$; $p < 0.05$; $\eta 2 = 0.35$]. Fisher LSD post hoc revealed that p1 in vertical was significantly different from p2 in vertical ($p < 0.01$), p3 in vertical ($p < 0.01$), p1 in diagonal ($p < 0.05$), and p3 in diagonal ($p < 0.05$), while all other comparisons were not significant (see **Figure 10B**). For the parietal ROI, none of the main effects were significant (all $p > 0.55$).

3.2 Source localization

After observing modulation of theta, alpha, and beta frequency, we observed the difference between diagonal and vertical movement in the generation of such frequency band activity in specific time windows (see Methods section for more details).

3.2.1 Theta

We analyzed theta during the peak in the forward period at the start of each trial, specifically between 100 and 300 ms. Significantly higher theta was found in the left middle frontal gyrus during diagonal movement compared to vertical movement ($p < 0.05$) (see **Figure 11**).

3.2.2 Alpha

We analyzed alpha in two time windows corresponding to the positive and negative activity peaks during the comeback period. These two time windows were P1 at 1200–1400 ms and P2 at 1650–1850 ms.

For P1, significant higher alpha was found in the posterior cingulate cortex during diagonal movement compared to vertical movement ($p < 0.05$). For P2, significantly higher alpha was found in the left inferior parietal lobule during vertical movement compared to diagonal movement ($p < 0.05$) (see **Figure 11B**).

3.2.3 Beta

We analyzed beta during the second time window in forward period (t2) and in the first time window (p1) during comeback period. These two time windows were t2 at 320–520 ms and p1 at 1220–1420 ms. None of the comparisons was significant (all $p > 0.52$).

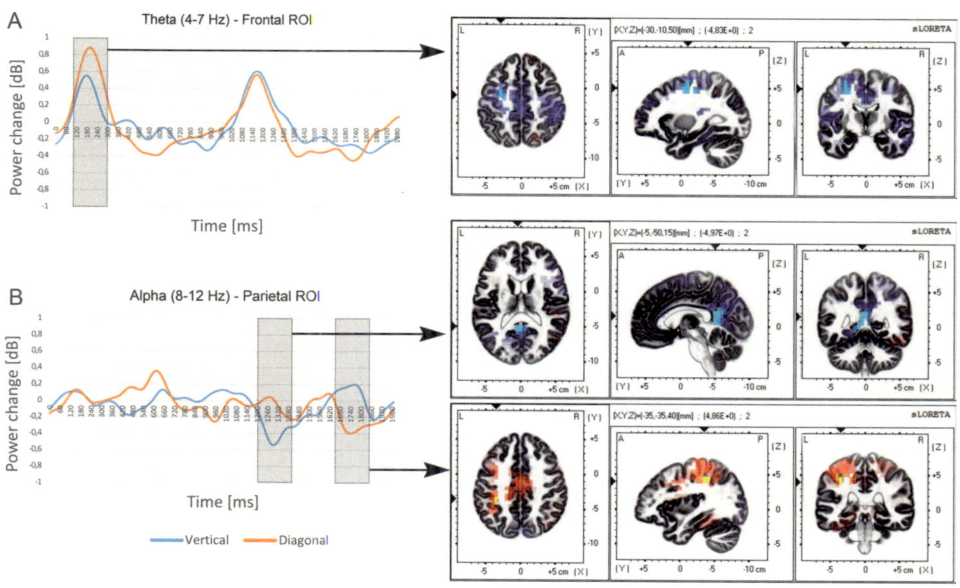

Figure 11.
Source localization of theta (A) and alpha (B) activity. Theta has been analyzed in the 100–300 ms time window. Alpha activity has been analyzed in two time windows: 1200–1400 and 1650–1850 ms (B). Blue blobs mean that activity during diagonal movement is higher than during vertical movement and vice versa for yellow blobs.

4. Discussion

Diagonal rhythmical bilateral movements have previously been suggested to promote improvement in motor and cognitive functions [1, 3]. The diagonal axis has the role of a metaphorical rule breaker in relation to the way of thinking, as suggested by the definition of "diagonal thinking" as a mixture of logical (i.e., vertical) and creative (i.e., lateral) thinking [47]. Diagonal movements are widely used in disciplines based on whole-body movements such as Asian martial arts (i.e., aikido, Tai Chi, Qigong) or modern and contemporary dance.

Yet, no study that we are aware of has actually examined electrophysiological changes during diagonal movements. Consequently, in the present study, we wanted to examine cerebral activity during diagonal arm movements in comparison to vertical arm movements. Previous literature has suggested that diagonal movements promote neuroplasticity along frontoparietal pathways through an increase in alpha and beta power [5]. For this reason, we focused our analysis on spectral power changes during the actual execution of movements in the theta, alpha, and beta bands.

Time-frequency analysis was conducted in order to observe how power in specific frequencies changes over time during movement execution.

4.1 Theta: attention, navigation, and computation

Comparing vertical to diagonal movement, we observed an increase of theta power during the first stage of diagonal movement over frontal electrodes, especially over Fz. This increment of frontal theta can be explained in three interrelated ways, including attentional effort [12, 19], navigational computation [13], and integration of information from different brain regions [16, 17]. First, it reflects greater attentional effort required for diagonal movement. In fact, diagonal movements are more complex and less automatic than vertical movements, and subjects must be more focused and pay more attention in order to perform correctly the movement. Second, it could reflect navigational computation. In both diagonal and vertical movements, the arms swing in peripersonal space, requiring the computation of the trajectory of both arms. However, it is only in diagonal movements that the arms go across the sagittal body midline and reach the other peripersonal hemispace. This kind of crossing requires more complex spatial computation than during movements that do not involve both hemispaces. Third, theta has previously been related to the integration of different information from distant cerebral regions [17]. Other studies have suggested that theta increase is related to a general mechanism of recall and integration of information from different domains supported by a central executive module [19, 48].

Finally, the state of focused attention on movement execution, together with the navigational computation necessary for the performance and the external sound, could evoke an absorption state in subjects. In fact, the role of *frontal midline theta* has been previously related to meditation, internalized attention, and integration of sensory information into executive control components of complex motor behavior [12, 15, 49].

4.2 Alpha: internalized attention and movement

While greater theta activity was related to the diagonal movements, we observed a biphasic modulation of alpha activity during the second part of vertical movements. Keeping in mind that each movement was composed of two parts (forward period and comeback period), we observed a biphasic response in the comeback period during vertical movement.

In fact, we observed first a decrease of alpha power immediately after the start of the comeback period and, consequently, an increase in the same frequency band tied to the end of the movement.

The decreased alpha activity observed immediately after the end of the second movement could reflect a decrease in internalized attention.

The decrease in alpha is classically related to a decrease of focusing on internal states and an amplified processing of information from the environment through sensorial inputs, especially from the visual system [17, 50].

Thus, this alpha decrease supports an increased attentional focus on the environment and a contextual decrease of internalized attention. Probably, after the forward period, subjects consider the movement ended, and this could represent an attentional disengagement related to the movement performance.

For those reasons, vertical movements in the comeback period do not require particular cognitive effort, but, immediately before the next trial, subjects must re-engage cognitive resources and focus on the internal state causing the significant increase of alpha observed at the end of the whole trial.

Klimesch [51] suggested that alpha is not a unitary frequency band but it would be better described as a union of two sub-bands named lower and upper alpha. These two sub-bands seem to support different functions: lower alpha would be more modulated by general attentional processes, while upper alpha would be more sensitive to perceptual and semantic processing during task performance.

Our result suggests that during vertical movement, Alpha1 and Alpha2 were significantly different in Peak2 and approached significance in Peak1. This means that during vertical movements, in Peak1, Alpha2 is slightly lower than Alpha1 and vice versa and in Peak2 Alpha2 is higher than Alpha1. Since Alpha2 has been associated with perceptual processing during task performance [51], we can hypothesize that the activity pattern observed over alpha power could reflect not a pure movement-related attentional engagement/disengagement but rather an inhibition/enhancement of environmental sensory input processing.

4.3 Beta

In the beta band, we observed a desynchronization during comeback period only in vertical movements, around 1300 ms after the start of the trial. The timing and the shape of the time course follow as observed in Peak1 in the alpha band. Contrary to our expectations, we did not find a similar desynchronization during diagonal movements. Probably, higher attentional demand during diagonal movement could address this issue, but further research will be performed.

Between consecutive trials (from 1900 to 100 ms of the next trial), we can observe a small increase in beta activity both in diagonal and vertical movements. This enhanced beta activity represents the post-movement beta rebound [35]. The presence of this activity suggests that subjects perceive the end of the trial as the conclusion of the movement, indexing the new trial as a new movement.

4.4 Source localization

A theta generator during the first stage of diagonal movements has been located in the middle frontal gyrus of the left hemisphere. The location of this generator supports the aforementioned frontal midline theta hypothesis in which the increase of theta during diagonal movement could increase internalized attention and integration of information from different regions such as frontal, sensorimotor, and parietal cortices.

Alpha generators during Peak1 and Peak2 in comeback period were found to be located in the posterior cingulate cortex and the left inferior parietal lobule, respectively. The posterior cingulate cortex is a region involved in directing attention toward internal or external cognition [52]. In our case, we observed a decrease of alpha during vertical movement. This result is in line with our hypothesis about the increase of external sensory input during Peak1 in vertical movement and the consequent inhibition in order to reallocate attention to the execution of the movement. In Peak2 we observed higher alpha activity during vertical compared to diagonal movement execution. The inferior parietal cortex is classically involved in spatial attention (for a review, see [53]), body perception [54], and motor functions [55]. Fu et al. [56] suggested that increase in alpha power over the inferior parietal cortex could underlie integration of sensory cues in order to redirect attention. This explanation fits well with our hypothesis of attentional fluctuations related to the end of vertical movements. In fact as already hypothesized before, the selective increase in alpha at the end of the movement could represent a re-engagement of attention toward sensorimotor information in order to perform correctly the next trial.

5. Conclusions

Results from time-frequency analysis and source localization converge to support our initial hypothesis. We observed that diagonal movements increase theta activity over the middle frontal gyrus during the first stage of movement execution, suggesting that they are indeed more complex than vertical movements. This complexity could be reflected in more computational resources being allocated at the start of the movement and to a reduced focus on the external environment during the whole performance. Moreover, the higher complexity of diagonal movement is addressed also by the biphasic alpha activity that involves the posterior cingulate cortex and inferior parietal lobule reflecting attentional fluctuations on internal vs. external environment during the execution of vertical movements.

Taken together, these results suggest that diagonal movements have indeed an effect on the ability to inhibit external input and to induce a mental mindset oriented to an increase in internalized attention.

Since theta has been related to a great variety of cognitive functions, one of our future aims is to understand whether motor training, which involves diagonal movement, could produce a reliable and prolonged enhancement of theta activity in order to enhance cognitive functioning on a long time scale.

We intend to examine whether the integrative role of theta could help healthy cognitive and motor development or be helpful in reducing symptoms of psychiatric disease based on psychological fragmentation, such as post-traumatic stress disorder or schizophrenia.

Conflict of interest

The authors declare that the research was conducted in the absence of any commercial or financial relationships that could be construed as a potential conflict of interest.

Author details

Fabio Marson[1], Patrizio Paoletti[1], Stefano Lasaponara[2], Joseph Glicksohn[3], Antonio De Fano[1] and Tal Dotan Ben-Soussan[1*]

1 Research Institute for Neuroscience, Education and Didactics, Patrizio Paoletti Foundation, Assisi, Italy

2 Libera Università Maria Santissima Assunta—LUMSA, Roma, Italy

3 Department of Criminology, and The Leslie and Susan Gonda (Goldschmied) Multidisciplinary Brain Research Center, Bar-Ilan University, Israel

*Address all correspondence to: research@fondazionepatriziopaoletti.org

References

[1] Diamond A. Effects of physical exercise on executive functions: Going beyond simply moving to moving with thought. Annals of sports medicine and research. 2015;**2**(1):1011

[2] Diamond A. Close interrelation of motor development and cognitive development and of the cerebellum and prefrontal cortex. Child Development. 2000;**71**(1):44-56

[3] Voss DE. Proprioceptive neuromuscular facilitation. American Journal of Physical Medicine & Rehabilitation. 1967;**46**(1):838-898

[4] Shimura K, Kasai T. Effects of proprioceptive neuromuscular facilitation on the initiation of voluntary movement and motor evoked potentials in upper limb muscles. Human Movement Science. 2002;**21**(1):101-113

[5] Moreira R, Lial L, Monteiro MGT, Aragão A, David LS, Coertjens M, et al. Diagonal movement of the upper limb produces greater adaptive plasticity than sagittal plane flexion in the shoulder. Neuroscience Letters. 2017;**643**:8-15

[6] Paoletti P, Glicksohn J, Ben-Soussan TD. Inner design technology: Improved affect by Quadrato motor training. In: The Amygdala—Where Emotions Shape Perception, Learning and Memories. Rijeka: InTech; 2017. pp. 27-41

[7] Ben-Soussan TD, Glicksohn J, Goldstein A, Berkovich-Ohana A, Donchin O. Into the square and out of the box: The effects of Quadrato motor training on creativity and alpha coherence. PLoS One. 2013;**8**(1):e55023

[8] Ben-Soussan TD, Glicksohn J, Berkovich-Ohana A. From cerebellar activation and connectivity to cognition: A review of the Quadrato motor training. BioMed Research International. 2015;**2015**:1-11

[9] Ben-Soussan TD, Glicksohn J, Berkovich-Ohana A, Donchin O, Goldstein A. Step in time: Changes in EEG coherence during a time estimation task following Quadrato motor training. Proceedings of Fechner Day. 2011;**27**(1):239-244

[10] Ben-Soussan TD, Berkovich-Ohana A, Glicksohn J, Goldstein A. A suspended act: Increased reflectivity and gender-dependent electrophysiological change following Quadrato motor training. Frontiers in Psychology. 2014;**5**:55

[11] Lasaponara S, Mauro F, Carducci F, Paoletti P, Tombini M, Quattrocchi CC, et al. Increased alpha band functional connectivity following the Quadrato motor training: A longitudinal study. Frontiers in Human Neuroscience. 2017;**11**:282

[12] Sauseng P, Hoppe J, Klimesch W, Gerloff C, Hummel FC. Dissociation of sustained attention from central executive functions: Local activity and interregional connectivity in the theta range. The European Journal of Neuroscience. 2007;**25**(2):587-593

[13] Buzsáki G, Moser EI. Memory, navigation and theta rhythm in the hippocampal-entorhinal system. Nature Neuroscience. 2013;**16**(2):130

[14] Klimesch W. EEG alpha and theta oscillations reflect cognitive and memory performance: A review and analysis. Brain Research Reviews. 1999;**29**(2-3):169-195

[15] Aftanas LI, Golocheikine SA. Human anterior and frontal midline theta and lower alpha reflect emotionally positive state and internalized attention: High-resolution EEG investigation of meditation. Neuroscience Letters. 2001;**310**(1):57-60

[16] Gruzelier J. A theory of alpha/theta neurofeedback, creative performance enhancement, long distance functional connectivity and psychological integration. Cognitive Processing. 2009;**10**(1):101-109

[17] Von Stein A, Sarnthein J. Different frequencies for different scales of cortical integration: From local gamma to long range alpha/theta synchronization. International Journal of Psychophysiology. 2000;**38**(3):301-313

[18] Nakashima K, Sato H. Relationship between frontal midline theta activity in EEG and concentration. Journal of Human Ergology. 1993;**22**(1):63-67

[19] Sauseng P, Griesmayr B, Freunberger R, Klimesch W. Control mechanisms in working memory: A possible function of EEG theta oscillations. Neuroscience and Biobehavioral Reviews. 2010;**34**(7):1015-1022

[20] Ben-Soussan TD, Mauro F, Lasaponara S, Glicksohn J, Marson F, Berkovich-Ohana A. Fully immersed: State absorption and electrophysiological effects of the OVO whole-body perceptual deprivation chamber. Progress in Brain Research. 2019;**244**:165-184

[21] Goldman RI, Stern JM, Engel J Jr, Cohen MS. Simultaneous EEG and fMRI of the alpha rhythm. Neuroreport. 2002;**13**(18):2487

[22] Pfurtscheller G, Stancak A Jr, Neuper C. Event-related synchronization (ERS) in the alpha band—An electrophysiological correlate of cortical idling: A review. International Journal of Psychophysiology. 1996;**24**(1-2):39-46

[23] Klimesch W, Doppelmayr M, Russegger H, Pachinger T, Schwaiger J. Induced alpha band power changes in the human EEG and attention. Neuroscience Letters. 1998;**244**(2):73-76

[24] Foxe JJ, Snyder AC. The role of alpha-band brain oscillations as a sensory suppression mechanism during selective attention. Frontiers in Psychology. 2011;**2**:154

[25] Feng W, Störmer VS, Martinez A, McDonald JJ, Hillyard SA. Involuntary orienting of attention to a sound desynchronizes the occipital alpha rhythm and improves visual perception. NeuroImage. 2017;**150**:318-328

[26] Lasaponara S, Pinto M, Aiello M, Tomaiuolo F, Doricchi F. The hemispheric distribution of α-band EEG activity during orienting of attention in patients with reduced awareness of the left side of space (spatial neglect). The Journal of Neuroscience. 2019;**39**(22):4332-4343

[27] Benedek M, Schickel RJ, Jauk E, Fink A, Neubauer AC. Alpha power increases in right parietal cortex reflects focused internal attention. Neuropsychologia. 2014;**56**:393-400

[28] Berger H. Über das elektrenkephalogramm des menschen. European Archives of Psychiatry and Clinical Neuroscience. 1929;**87**(1):527-570

[29] Harland CJ, Clark TD, Prance RJ. Remote detection of human electroencephalograms using ultrahigh input impedance electric potential sensors. Applied Physics Letters. 2002;**81**(17):3284-3286

[30] Buzsaki G. Rhythms of the Brain. New York: Oxford University Press; 2006

[31] Pfurtscheller G, Da Silva FL. Event-related EEG/MEG synchronization and desynchronization: Basic principles. Clinical Neurophysiology. 1999;**110**(11):1842-1857

[32] Kilavik BE, Zaepffel M, Brovelli A, MacKay WA, Riehle A. The ups and downs of beta oscillations in sensorimotor cortex. Experimental Neurology. 2013;**245**:15-26

[33] Neuper C, Wörtz M, Pfurtscheller G. ERD/ERS patterns reflecting sensorimotor activation and deactivation. Progress in Brain Research. 2006;**159**:211-222

[34] Pfurtscheller G, Stancak A Jr, Neuper C. Post-movement beta synchronization. A correlate of an idling motor area? Electroencephalography and Clinical Neurophysiology. 1996;**98**(4):281-293

[35] Parkes LM, Bastiaansen MC, Norris DG. Combining EEG and fMRI to investigate the post-movement beta rebound. NeuroImage. 2006;**29**(3):685-696

[36] Ray WJ, Cole HW. EEG alpha activity reflects attentional demands, and beta activity reflects emotional and cognitive processes. Science. 1985;**228**(4700):750-752

[37] Papanicolaou AC, Loring DW, Deutsch G, Eisenberg HM. Task-related EEG asymmetries: A comparison of alpha blocking and beta enhancement. The International Journal of Neuroscience. 1986;**30**(1-2):81-85

[38] Egner T, Gruzelier JH. EEG biofeedback of low beta band components: Frequency-specific effects on variables of attention and event-related brain potentials. Clinical Neurophysiology. 2004;**115**(1):131-139

[39] Gola M, Magnuski M, Szumska I, Wróbel A. EEG beta band activity is related to attention and attentional deficits in the visual performance of elderly subjects. International Journal of Psychophysiology. 2013;**89**(3):334-341

[40] Delorme A, Makeig S. EEGLAB: An open source toolbox for analysis of single-trial EEG dynamics including independent component analysis. Journal of Neuroscience Methods. 2004;**134**(1):9-21

[41] Onton J, Makeig S. Information-based modeling of event-related brain dynamics. Progress in Brain Research. 2006;**159**:99-120

[42] Cohen MX, Elger CE, Ranganath C. Reward expectation modulates feedback-related negativity and EEG spectra. NeuroImage. 2007;**35**(2):968-978

[43] Pascual-Marqui RD, Esslen M, Kochi K, Lehmann D. Functional imaging with low-resolution brain electromagnetic tomography (LORETA): A review. Methods and Findings in Experimental and Clinical Pharmacology. 2002;**24**(Suppl C):91-95

[44] Nichols TE, Holmes AP. Nonparametric permutation tests for functional neuroimaging: A primer with examples. Human Brain Mapping. 2002;**15**(1):1-25

[45] Missonnier P, Hasler R, Perroud N, Herrmann FR, Millet P, Richiardi J, et al. EEG anomalies in adult ADHD subjects performing a working memory task. Neuroscience. 2013;**241**:135-146

[46] Maclin EL, Mathewson KE, Low KA, Boot WR, Kramer AF, Fabiani M, et al. Learning to multitask: Effects of video game practice on electrophysiological indices of attention and resource allocation. Psychophysiology. 2011;**48**(9):1173-1183

[47] De Bono E. Information processing and new ideas—Lateral and vertical thinking. The Journal of Creative Behavior. 1969;**3**(3):159-171

[48] Sauseng P, Klimesch W, Schabus M, Doppelmayr M. Fronto-parietal EEG

coherence in theta and upper alpha reflect central executive functions of working memory. International Journal of Psychophysiology. 2005;**57**(2):97-103

[49] Kubota Y, Sato W, Toichi M, Murai T, Okada T, Hayashi A, et al. Frontal midline theta rhythm is correlated with cardiac autonomic activities during the performance of an attention demanding meditation procedure. Cognitive Brain Research. 2001;**11**(2):281-287

[50] Adrian ED. The Physical Background of Perception. Mind. 1948;**226**:244-249

[51] Klimesch W, Doppelmayr M, Pachinger T, Russegger H. Event-related desynchronization in the alpha band and the processing of semantic information. Cognitive Brain Research. 1997;**6**(2):83-94

[52] Leech R, Sharp DJ. The role of the posterior cingulate cortex in cognition and disease. Brain. 2013;**137**(1):12-32

[53] Behrmann M, Geng JJ, Shomstein S. Parietal cortex and attention. Current Opinion in Neurobiology. 2004;**14**(2):212-217

[54] Daprati E, Sirigu A, Nico D. Body and movement: Consciousness in the parietal lobes. Neuropsychologia. 2010;**48**(3):756-762

[55] Fogassi L, Luppino G. Motor functions of the parietal lobe. Current Opinion in Neurobiology. 2005;**15**(6):626-631

[56] Fu KMG, Foxe JJ, Murray MM, Higgins BA, Javitt DC, Schroeder CE. Attention-dependent suppression of distracter visual input can be cross-modally cued as indexed by anticipatory parieto–occipital alpha-band oscillations. Cognitive Brain Research. 2001;**12**(1):145-152

Chapter 3

Correlations of Gait Phase Kinematics and Cortical EEG: Modelling Human Gait with Data from Sensors

Chaitanya Nutakki, Sandeep Bodda and Shyam Diwakar

Abstract

Neural coding of gait intent and continuous gait kinematics have advanced brain computer interface (BCI) technology for detection and predicting human upright walking movement. However, the dynamics of cortical involvement in upright walking and upright standing has not been clearly understood especially with the focus of off-laboratory assessments. In this study, wearable low-cost mobile phone accelerometers were used to extract position and velocity at 12 joints during walking and the cortical changes involved during gait phases of walking were explored using non-invasive electroencephalogram (EEG). Extracted gait data included, accelerometer values proximal to brachium of arm, antecubitis, carpus, coxal, femur and tarsus by considering physical parameters including height, weight and stride length. Including EEG data as features, the spectral and temporal features were used to classify and predict the swing and stance instances for healthy subjects. While focusing on stance and swing classification in healthy subjects, this chapter relates to gait features that help discriminate walking movement and its neurophysiological counterparts. With promising initial results, further exploration of gait may help change detection of movement neurological conditions in regions where specialists and clinical facilities may not be at par.

Keywords: human gait, cortical activation, electroencephalography, stance, swing, accelerometer sensors

1. Introduction

Upright gait has been used as a peculiar biometric characteristic and can offer clues to help develop detection mechanisms for walking-related neurological disorders, if detected can help reduce cost and help propose diagnostic approaches [1]. Gait and locomotion are complex sequential processes involving timed coordination between central nervous system, muscles and bones [2]. The action of numerous muscles and the variability in joint kinematics, leads to changes between different phases in gait, mainly swing and stance [3]. Human gait analysis involves the measurement and assessment of kinematic and inverse dynamic parameters that characterize the different phases of gait and quantifies the musculoskeletal functions [4, 5]. Within the context of gait measurement, sensors used in assessing

GAIT include floor based sensors (FS), wearable sensors (WS) and non-wearable sensors (NWS) techniques [6]. In techniques involving NWS, gait has been captured using cameras with video and image processing allowing features to attribute to gait patterns [7]. FS based techniques use force plates located on the ground to extract the walking parameters through pressure estimates and ground reaction forces [8]. In WS based techniques, sensors like accelerometers [9–11], gyroscopes, goniometers, etc. attached to the different biomechanical parts of the human body frame measure gait patterns during walking [12]. Peak detection methods in algorithms allow gauging gait events like heel strike, swing and stance from accelerometer and gyroscope data [13–16].

Aiming towards potential applications in medicine, it has been noted that gait and posture control in patients with neurodegenerative disorders become irregular due to weakening of motor neurons that controls the muscles [17]. Neurodegenerative diseases including the Parkinson's and Huntington disease result in progressive degeneration of neurons causes changes in neuromuscular control [18]. In clinical analysis, in order to understand the patient's walking capability and movement tracking usually require expensive (cost, effort and time) methodologies and structured laboratories [19, 20]. A study by Hausdorff et al. [21] had demonstrated the differences between gait cycles and subphases duration in Parkinson's patients compared to normal subjects. Also, magnitude difference between gait stride intervals of human subjects with neurodegenerative conditions have been analysed by using (DFAT) detrended fluctuation analysis techniques [22].

Research progress in understanding the brain function during gait intent, but the information on movement-related cortical activity, neural circuit mechanisms and computations underlying the control of upright walking in humans are yet to be understood completely [23, 24]. Studies have shown rhythmic foot and leg movements recruit primary motor cortex [25, 26], while fNIRS has shown involvement of frontal, premotor and supplementary motor areas during walking [27, 28]. Recent literature have indicated augmented beta oscillations during double support phases of the gait cycle (event-related synchronization, ERS) and to be suppressed during the swing and single support phases (event-related desynchronization, ERD) [29–35]. Other studies have shown enhanced gamma oscillations during early and mid-swing phases of gait cycle and suppressed gamma rhythms towards the end of the swing phase and during the double support [36–40].

Studies involving other techniques such as single-photon emission computed tomography (SPECT) have reported neural characteristics during voluntary walking; studies [41, 42] using SPECT evaluated changes in the brain activity as a result of walking, and identified the SMA, S1, M1, cerebellum, and basal ganglia functioning as the control mechanisms of bipedal gait. Another SPECT study [43], investigated cortical activation during treadmill walking and found network activation in the premotor cortex, somatosensory association cortex, cingulate cortex and brain stem apart from the structures reported [41]. In both tomography SPECT studies, walking tasks were carried out prior to image acquisition.

Additionally, neurodegenerative diseases that relate to gait effects can be classified using machine learning tools as a decision support to clinicians for better prediction [44] of patient conditions. Gait disorder related to amyotrophic lateral sclerosis (ALS) have been classified by using wavelet-based scheme and features reflect regularity and gait coherence between both limbs as seen from the approximation part of the raw gait signal [45].

Identification and classification of human gait using low cost experimental techniques are crucial and necessary for the developing countries like India. Current diagnosing techniques for gait related disorders are more expensive and inaccessible to the common people. The main purpose of this study was to develop a low-cost model that can be employed in the future as an off-laboratory diagnosing tool for identifying gait abnormalities and classify human gait phases and pathological disorders. With machine learning jointly with inverse dynamic analysis using triaxial accelerometer sensors in today's mobile phones, it may be reliable to analyse the each joint kinematic and behaviour during swing and stance phases that can further be used for the development of control strategies to set up brain machine interfaces (BMI), human-machine interfaces (HMI) and prostheses. Gait kinematic movement in terms of EEG allowed to understand the cortical regions that are active during gait phases helps in diagnosing the gait neurological disorders.

In this chapter, we address the usage of low-cost mobile phone-based accelerometer sensors in order to extract and analyse human gait patterns. Average torque and the gait kinematic parameters of the lower body during stance and swing phases were analysed to understand how gait can be attained. In this study, we compared neural spectral representations from scalp EEG signals during active walking. Simultaneous recording of EEG with gait and their analysis was done to interpret cortical activity during the stance and swing phases of a gait cycle.

2. Methods

2.1 Low cost sensor-based gait recording and assessment

Gait data was extracted from 20 healthy volunteers using 12 smartphone-based accelerometers and a software application that allowed synchronous collection of data from the devices and mapped to additional parameters, including weight and age. The data collection and methods were approved by the institutional ethical review board and an open consent was collected from the participants prior to gait and EEG recordings. A total of 40 trails and 120 gait cycle accelerometer data were extracted from brachium of arm (shoulder), antecubitis (elbow), carpus (wrist), coxal (hip), femur (knee) and tarsus (ankle) were taken for further analysis. The extracted data was then normalized and sixth order Butterworth filter with a cut-off frequency of 10 Hz was used for noise reduction. Data processing was based on the time noted by the observer during each gait phase (**Figure 1A**).

2.2 Estimating torque amplitude for each joint

This method employed 12 joint related positions to collect data from subjects. Joint torques (Eq. (1)) were calculated by providing joint length, force and angle to compute muscle force that attributed to joint rotation at different gait phases.

$$T_{j..n} = Fi * R * \sin\theta \tag{1}$$

$$F_{i..n} = m_{i..n} * a_{i..n} \tag{2}$$

Here, '$F_i...n$' was the force (Eq. (2)) of each joint (i) derived from mass and acceleration of each joint, where the acceleration and angle were directly retrieved from the accelerometer sensor, 'R' was taken as length of joint measured before the experiment was done.

Average torque amplitude was computed as A, the average torque amplitude for each joint (Eq. (3)).

$$A_j = \frac{1}{T}\sum_{t=1}^{T} A_j(t) \tag{3}$$

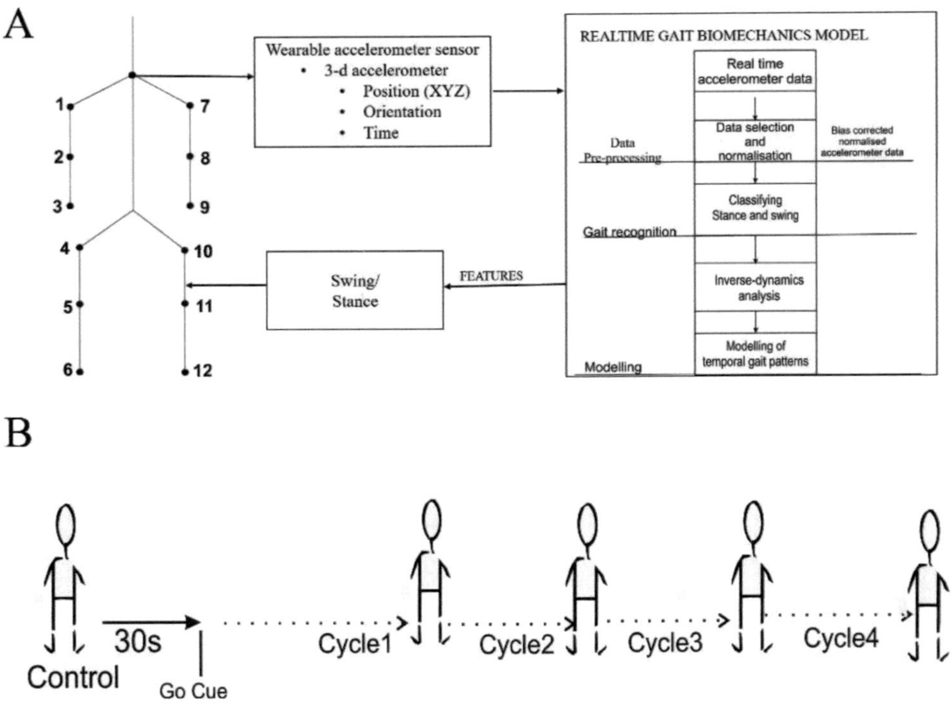

Figure 1.
(A) Classifying swing stance from gait data. (B) Schematic representation of EEG recording protocol for gait.

2.3 Experimental recording of gait using EEG

EEG was measured from four healthy subjects with four cycles per trial and two trials were recorded per subject. All subjects had provided their informed consent and were approved by the institutional ethical review board. Subjects were without any known medical conditions and had normal or corrected-to-normal vision. All subjects were explained the aim of the study before participating in the recordings. Contiguous EEG was recorded (**Figure 1B**) from 14 Ag/AgCl electrodes positioned on a commercial scalp cap based on the 10–20 placement system at a sampling rate of 128 Hz using electrode (see **Figure 1B** for protocol).

All trials began with a relaxation phase (blank screen), considered as reference or baseline signal for the analysis; 30 seconds of the relaxation phase was followed by upright active walking after the audio cue 'START'. Subject performed four cycles with each cycle including a swing and a stance phases starting with right leg followed by left leg.

Signal processing and data analysis on raw EEG signal were performed using custom scripts MATLAB R2017b (MathWorks, Massachusetts, USA) and EEGLAB toolbox [46] The reference channel (mean) was subtracted, detrended and band-pass filtered using a FIR filter of order 20 within the range, 0.1–60 Hz and notch filter applied to remove line noise of 50 Hz. Based on the marker points for various tasks (relax, step1, step 2,..., step 8, stop), data was extracted for each task and the power spectrum was estimated for the relative bands α/μ, β and γ in the EEG signals. Spectral bands were estimated for each stance and swing phases of gait cycle and averaged across all the trials of the four healthy subjects. From the pre-processed epochs, estimated bands were quantified and significant regions were identified.

3. Results

3.1 Torque variations across the subjects define swing and stance gait phases

Joint torques during stance and swing phases were measured using inverse dynamic analysis to understand how gait can be attained [47] at different phases. Subjects were divided into groups with respect to weights. Group I was categorized in to subjects with 60–70 kg, Group II was categorized in to subject with 70–80 kg, Group III was categorized into subjects with 80–90 kg. Average torque amplitude of subjects was analysed and compared during swing (**Figure 2A**) and stance (**Figure 2B**) phases.

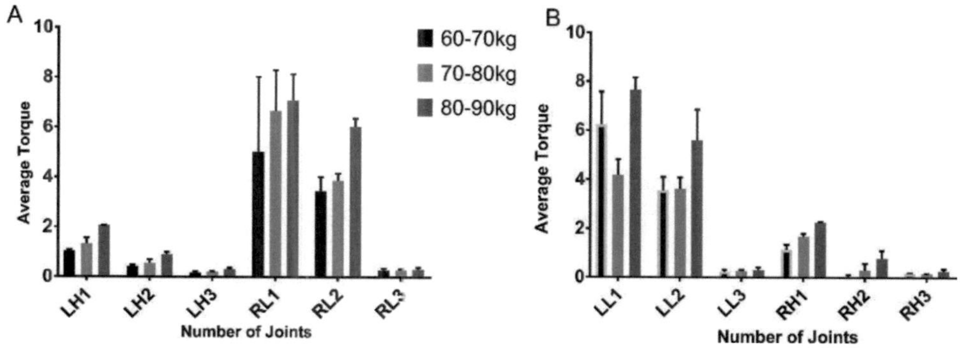

Figure 2.
Torque variations during gait. (A) Average torque amplitude of joints with different weight groups during swing. (B) Average torque amplitude of joints with different weight groups during stance. RH, right hip; RK, right knee; RA, right ankle; LS, left shoulder; LE, left elbow; LW, left wrist; LH, left hip; LK, left knee; RS, right shoulder; RE, right elbow; RW, right wrist and 1–3 are joint positions.

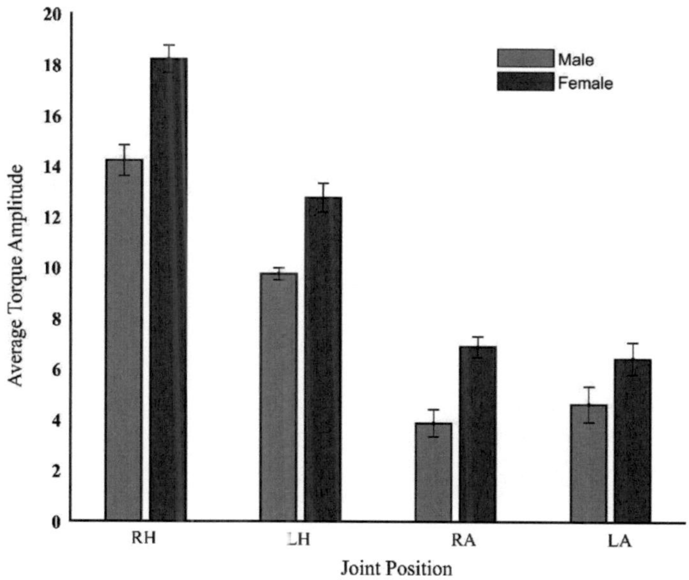

Figure 3.
Torque amplitude changes across hip and ankle of male and female subjects. Hip and ankle joints show significant changes allowing classification of male and female subjects. RH, right hip; LH, left hip; RA, right ankle; LA, left ankle.

3.2 Variations in torque allow to differentiate male and female gait

Lower body torque amplitude of male and female were analysed and compared during swing and stance from frontal plane. Torque amplitude of hip and ankle of female joints in the frontal plane showed more activity than the male in frontal plane (**Figure 3**).

3.3 Temporal and spectral EEG features of gait

Gait-related cortical potentials include the positive and negative motor potentials at the onset of movement for swing and stance. Positive amplitude of motor potential has observed for swing phase of the gait cycle in the frontal electrodes (F3) whereas negative amplitude of motor potential has observed for stance phase of the gait cycle in the frontal electrodes (F3). The clear distinction of motor potential has shown (**Figure 4A** and **B**).

From the spectral maps over the comparison of swing (**Figure 5C**) and stance (**Figure 5D**) and we have observed higher activity in parietal and frontal regions over the low frequency band regions delta and theta bands. Also, decreased alpha and beta band in frontal and central cortical regions were observed during swing than during stance phase. However, only right swing and left stance were explored in this study.

3.4 Classifying gait sensorial data using different machine learning algorithms

Since gait cadence has nonlinear and complex behaviours, extracted gait data was classified using different machine learning algorithms with validations using percentage split (60 and 70%) methods. Training accuracies suggest most algorithms had similar Among all the tested algorithms [48], Naïve Bayes and SVM with linear kernel showed highest training accuracies as in other studies [44, 49, 50] across different splits with gait accelerometer data (see **Figure 6**). We also tried leave-one-out-cross-validation but had similar results (data not shown). The data suggests that machine learning methods may help predict normal gait phases with torque features. Although recorded simultaneously, since EEG classification using machine learning was not done in this study, we may need to explore a potential technique for identifying gait phases in terms of spectral compositions. Errors were attributed to variability in data from accelerometer time and frequency fluctuations due to different models used (data not shown).

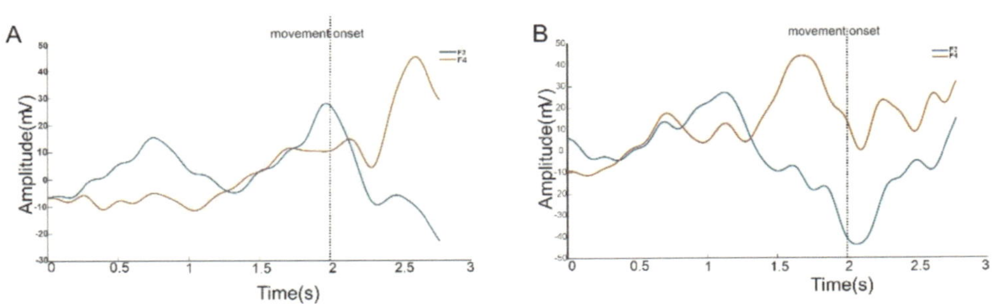

Figure 4.
Gait related cortical potentials: evoked average response for swing and stance phase of gait cycle (A) time course of F3 (blue) F4 (red) response of swing phase of gait cycle showing positive amplitude at the movement onset (B) time course of F3 (blue), F4 (red) response of stance phase of gait cycle shows negative amplitude at the movement onset.

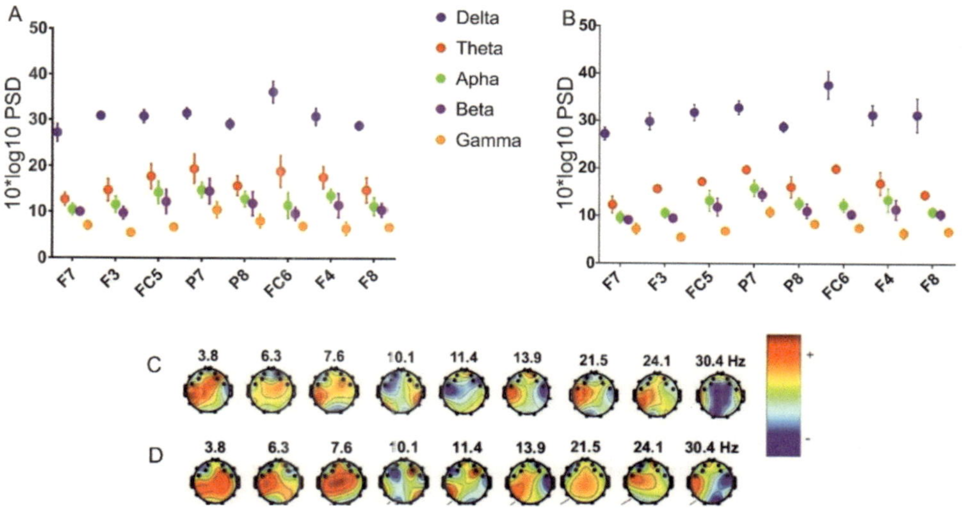

Figure 5.
Spectral changes for swing and stance phases of a gait cycle. (A) Swing phase of gait cycle showing higher delta and theta bands in frontal regions (F8, F3 and F7) electrodes. (B) Stance phase of gait cycle showing higher delta band parietal regions (P7 and P8). (C and D) Scalp maps for frequency ranges during swing (C) and stance gait phases (D).

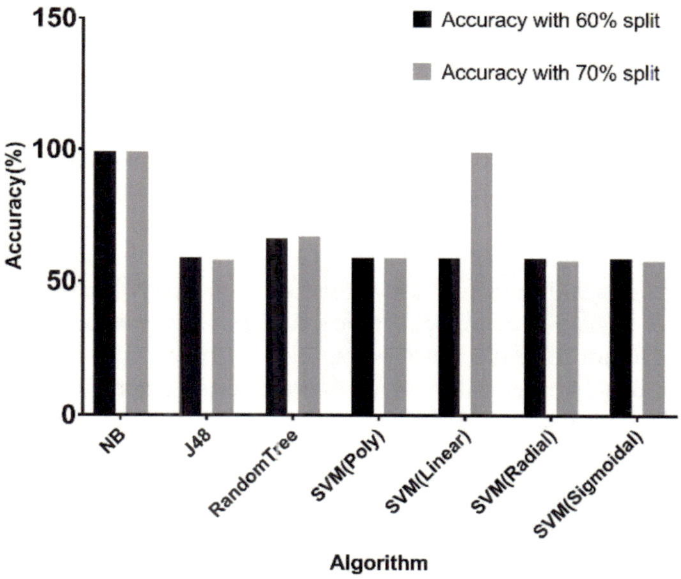

Figure 6.
Classification of gait data using machine learning algorithms. Naïve Bayes (NB), J48 decision tree, random tree, support vector machine algorithms with polynomial, linear, radial and sigmoidal and radial basis functions allowed classifying gait data.

Support vector machines had 58–59% training accuracies and so did J48 algorithm. Random tree had 67% while Naïve Bayes and linear SVM showed more than 98% accuracy perhaps attributed to complex decision boundaries.

4. Discussion

Torque-based reconstructions of gait from mobile phone triaxial accelerometer data may help identifying swing and stance phases in gait in addition to allowing specific joint based data for discriminating male and female characteristics in gait. Reliably using EEG to predict swing and stance will include comparisons of temporal and spectral components although the resolutions and accuracies are not so reliable beyond basic gait changes, we find the positive and negative amplitudes of the MRCPs can serve as good discriminators.

Gait data was classified using machine learning algorithms with percentage split cross-validations. As with many datasets, with increase in training data samples, a consequential increase in the accuracy was observed. Among the algorithms Naïve Bayes, SVM and tree-based algorithms showed high accuracy across the data with validations based on different percentage splits of training data. The data from accelerometers may be used in the BCI-related predictive algorithms for gait phase estimates.

The study computed joint torques in order to understand relationship of joint rotations during gait phases. As indicated, generated torque amplitude was sufficient to test classification algorithms on accelerometer-based gait data. We analysed the data grouped based on the subject weight since average torque amplitude of each subject was dependent on the weight of the subject. As the weight of the subject increased, increments in the joint torques were observed across the subjects. The torques and forces within subjects during different gait cycles showed little difference.

In terms of gait data from accelerometers, male subjects showed variations in the frontal and sagittal axes and estimates suggested higher joint movement correlated to higher torque amplitude changes with respect to motion. Hip and ankle joints served as strong discriminators in classification of subject gender based on data. Rather than acceleration, torques classified variations of gait across male and female subjects.

EEG-gait methodology allowed to map cortical organization relationships and between the contralateral and ipsilateral joints during gait. During stance when compared to swing, there was higher activity in the delta and theta bands in the frontal and parietal regions, whereas decreased activity in beta band in the parietal regions. Using delta and beta rhythms in the fronto-parietal cortical microzones, it may be possible to classify swing and stance. Additionally, gait-based assessments need to rely on motor related cortical potentials and their amplitudes. Temporal analysis of gait related potentials has shown positive and negative motor potentials for stance and swing and their significant variety could be related as a marker discriminating stance and swing.

The significance of such assessments is many; with gait categorization using torque, it may now be possible to employ mobile phone accelerometers to estimate swing and stance variations as a preclinical step for estimating medical disorders. The variations could also allow gait as a biometric information especially in validating male and female subjects and their upright walking capabilities. Although EEG data is far from assessing gait intent, initialization, swing and stance phases may be explored for correlations related to neurophysiological changes attributing such data for classifying neurological disorders in the future.

5. Conclusions

Spatio-temporal reconstruction of swing and stance from triaxial accelerometers allow an understanding of how multi-position accelerometer data accounts for healthy gait before developing optimizations and methods to assess dysfunctional gait. The study suggests quantifying specific torque patterns during gait may facilitate cheaply and easily detecting gait phase changes. Although a more detailed and multi-configuration data relating gait and EEG may be needed, this step helps to propose a pre-clinical assessment tool for rural communities, especially when multi-specialty hospitals may consider outreach or where specialists may need more time to understand movement related conditions prior to an actual diagnosis. With useful preliminary results that supports gait as a BCI technology, it further warrants the need to investigate the utility of mobile phone sensors for extracting accelerometer-based data and its use in a patient population.

Acknowledgements

This work derives direction and ideas from the Chancellor of Amrita University, Sri Mata Amritanandamayi Devi. This work was partially supported by Grants DST/CSRI/2017/31 of Department of Science and Technology, Visvesvaraya Fellowship, Ministry of Electronics and IT, Government of India and by Embracing The World, Research-for-a-Cause-Initiative.

Conflict of interest

The authors declare no conflict of interest.

Author details

Chaitanya Nutakki, Sandeep Bodda and Shyam Diwakar*
Amrita School of Biotechnology, Amrita Vishwa Vidyapeetham, Amritapuri Campus, Kollam, Kerala, India

*Address all correspondence to: shyam@amrita.edu

References

[1] Nonnekes J, Goselink RJM, Růžička E, Fasano A, Nutt JG, Bloem BR. Neurological disorders of gait, balance and posture: A sign-based approach. Nature Reviews Neurology. 2018;**14**:183-189

[2] Pedotti A. A study of motor coordination and neuromuscular activities in human locomotion. Biological Cybernetics. 1977;**26**:53-62

[3] Stickford ASL, Stickford JL. Ventilation and locomotion in humans: Mechanisms, implications, and perturbations to the coupling of these two rhythms. Springer Science Reviews. 2014;**2**:95-118

[4] Novacheck TF. The biomechanics of running. Gait & Posture. 1998;**7**:77-95

[5] Cappellini G, Ivanenko YP, Poppele RE, Lacquaniti F. Motor patterns in human walking and running. Journal of Neurophysiology. 2006;**95**:3426-3437

[6] Gafurov D. A Survey of Biometric Gait Recognition: Approaches, Security and Challenges. Proceedings of the Norwegian Informatics Conference; 2007. pp. 19-21

[7] Baratin E, Sugavaneswaran L, Umapathy K, Ioana C, Krishnan S. Wavelet-based characterization of gait signal for neurological abnormalities. Gait & Posture. 2015;**41**:634-639

[8] Mostayed A, Kim S, Mazumder MMG, Park SJ. Foot step based person identification using histogram similarity and wavelet decomposition. In: 2008 Int. Conf. Inf. Secur. Assur. (isa 2008); IEEE; 2008. pp. 307-311

[9] Godfrey A, Del Din S, Barry G, Mathers JC, Rochester L. Instrumenting gait with an accelerometer: A system and algorithm examination. Medical Engineering & Physics. 2015;**37**:400-407

[10] Patterson M, Caulfield B. A novel approach for assessing gait using foot mounted accelerometers. In: Proc 5th Int ICST Conf Pervasive Comput Technol Healthc; 2011. DOI: 10.4108/icst.pervasivehealth.2011.246061

[11] Del Din S, Hickey A, Hurwitz N, Mathers JC, Rochester L, Godfrey A. Measuring gait with an accelerometer-based wearable: Influence of device location, testing protocol and age. Physiological Measurement. 2016;**37**:1785-1797

[12] Mantyjarvi J, Lindholm M, Vildjiounaite E, Makela S, Ailisto H. Identifying users of portable devices from gait pattern with accelerometers. In: Proceedings. (ICASSP '05). IEEE Int. Conf. Acoust. Speech, Signal Process. IEEE; 2005. pp. 973-976

[13] Selles RW, Formanoy MAG, Bussmann JBJ, Janssens PJ, Stam HJ. Automated estimation of initial and terminal contact timing using accelerometers; development and validation in transtibial amputees and controls. IEEE Transactions on Neural Systems and Rehabilitation Engineering. 2005;**13**:81-88

[14] Derawi MO, Bours P, Holien K. Improved cycle detection for accelerometer based gait authentication. In: 2010 Sixth Int. Conf. Intell. Inf. Hiding Multimed. Signal Process. IEEE; 2010. pp. 312-317

[15] Casamassima F, Ferrari A, Milosevic B, Ginis P, Farella E, Rocchi L. A wearable system for gait training in subjects with Parkinson's disease. Sensors. 2014;**14**:6229-6246

[16] Rebula JR, Ojeda LV, Adamczyk PG, Kuo AD. Measurement of foot placement and its variability with inertial sensors. Gait & Posture. 2013;**38**:974-980

[17] Wu Y, Krishnan S. Computer-aided analysis of gait rhythm fluctuations in amyotrophic lateral sclerosis. Medical & Biological Engineering & Computing. 2009;**47**:1165-1171

[18] Schlachetzki JCM, Barth J, Marxreiter F, et al. Wearable sensors objectively measure gait parameters in Parkinson's disease. PLoS ONE. 2017;**12**:e0183989

[19] Komaris D-S, Govind C, Murphy A, Ewen A, Riches P. Identification of movement strategies during the sit-to-walk movement in patients with knee osteoarthritis. Journal of Applied Biomechanics. 2018;**34**:96-103

[20] Barth J, Klucken J, Kugler P, Kammerer T, Steidl R, Winkler J, et al. Biometric and mobile gait analysis for early diagnosis and therapy monitoring in Parkinson's disease. In: 2011 Annu. Int. Conf. IEEE Eng. Med. Biol. Soc. IEEE; 2011. pp. 868-871

[21] Hausdorff JM, Lertratanakul A, Cudkowicz ME, Peterson AL, Kaliton D, Goldberger AL. Dynamic markers of altered gait rhythm in amyotrophic lateral sclerosis. Journal of Applied Physiology. 2000;**88**:2045-2053

[22] Hausdorff JM, Cudkowicz ME, Firtion R, Wei JY, Goldberger AL. Gait variability and basal ganglia disorders: Stride-to-stride variations of gait cycle timing in parkinson's disease and huntington's disease. Movement Disorders. 1998;**13**:428-437

[23] Presacco A, Goodman R, Forrester L, Contreras-Vidal JL. Neural decoding of treadmill walking from noninvasive electroencephalographic signals. Journal of Neurophysiology. 2011;**106**:1875-1887

[24] Choi JT, Bastian AJ. Adaptation reveals independent control networks for human walking. Nature Neuroscience. 2007;**10**:1055-1062

[25] Sburlea AI, Montesano L, Cano-De La Cuerda R, Alguacil Diego IM, Miangolarra-Page JC, Minguez J. Detecting intention to walk in stroke patients from pre-movement EEG correlates. Journal of Neuroengineering and Rehabilitation. 2015;**12**:1-12

[26] Hortal E, Úbeda A, Iáñez E, Fernández E, Azorín JM. Using EEG Signals to Detect the Intention of Walking Initiation and Stop. Cham: Springer; 2015. pp. 278-287

[27] Perrey S. Possibilities for examining the neural control of gait in humans with fNIRS. Frontiers in Physiology. 2014;**5**:204

[28] Jin H, Li C, Xu J. Pilot study on gait classification using fNIRS signals. Computational Intelligence and Neuroscience. 2018;**2018**:1-9

[29] Artoni F, Fanciullacci C, Bertolucci F, Panarese A, Makeig S, Micera S, et al. Unidirectional brain to muscle connectivity reveals motor cortex control of leg muscles during stereotyped walking. NeuroImage. 2017;**159**:403-416

[30] Bradford JC, Lukos JR, Ferris DP. Electrocortical activity distinguishes between uphill and level walking in humans. Journal of Neurophysiology. 2016;**115**:958-966

[31] Bruijn SM, Van Dieën JH, Daffertshofer A. Beta activity in the premotor cortex is increased during stabilized as compared to normal walking. Frontiers in Human Neuroscience. 2015;**9**:593

[32] Bulea TC, Kim J, Damiano DL, Stanley CJ, Park H-S. Prefrontal, posterior parietal and sensorimotor network activity underlying speed control during walking. Frontiers in Human Neuroscience. 2015;**9**:247

[33] Cheron G, Duvinage M, De Saedeleer C, et al. From spinal central pattern generators to cortical network: Integrated BCI for walking rehabilitation. Neural Plasticity. 2012;**2012**:375148

[34] Gwin JT, Gramann K, Makeig S, Ferris DP. Electrocortical activity is coupled to gait cycle phase during treadmill walking. NeuroImage. 2011;**54**:1289-1296

[35] Knaepen K, Mierau A, Swinnen E, Fernandez Tellez H, Michielsen M, Kerckhofs E, et al. Human-robot interaction: Does robotic guidance force affect gait-related brain dynamics during robot-assisted treadmill walking? PLoS ONE. 2015;**10**:e0140626

[36] Seeber M, Scherer R, Wagner J, Solis-Escalante T, Müller-Putz GR. EEG beta suppression and low gamma modulation are different elements of human upright walking. Frontiers in Human Neuroscience. 2014;**8**:485

[37] Seeber M, Scherer R, Wagner J, Solis-Escalante T, Müller-Putz GR. High and low gamma EEG oscillations in central sensorimotor areas are conversely modulated during the human gait cycle. NeuroImage. 2015;**112**:318-326

[38] Storzer L, Butz M, Hirschmann J, Abbasi O, Gratkowski M, Saupe D, et al. Bicycling and walking are associated with different cortical oscillatory dynamics. Frontiers in Human Neuroscience. 2016;**10**:61

[39] Wagner J, Solis-Escalante T, Grieshofer P, Neuper C, Müller-Putz G, Scherer R. Level of participation in robotic-assisted treadmill walking modulates midline sensorimotor EEG rhythms in able-bodied subjects. NeuroImage. 2012;**63**:1203-1211

[40] Wagner J, Solis-Escalante T, Scherer R, Neuper C, Müller-Putz G. It's how you get there: Walking down a virtual alley activates premotor and parietal areas. Frontiers in Human Neuroscience. 2014;**8**:93

[41] Fukuyama H, Ouchi Y, Matsuzaki S, Nagahama Y, Yamauchi H, Ogawa M, et al. Brain functional activity during gait in normal subjects: A SPECT study. Neuroscience Letters. 1997;**228**:183-186

[42] Wieser M, Haefeli J, Bütler L, Jäncke L, Riener R, Koeneke S. Temporal and spatial patterns of cortical activation during assisted lower limb movement. Experimental Brain Research. 2010;**203**:181-191

[43] Hanakawa T, Katsumi Y, Fukuyama H, Honda M, Hayashi T, Kimura J, et al. Mechanisms underlying gait disturbance in Parkinson's disease. Brain. 1999;**122**:1271-1282

[44] Nutakki C, Narayanan J, Anchuthengil AA, Nair B, Diwakar S. Classifying gait features for stance and swing using machine learning. In: 2017 Int Conf Adv Comput Commun Informatics, ICACCI 2017; 2017. DOI: 10.1109/ICACCI.2017.8125896

[45] Wu Y, Shi L. Analysis of altered gait cycle duration in amyotrophic lateral sclerosis based on nonparametric probability density function estimation. Medical Engineering & Physics. 2011;**33**:347-355

[46] Delorme A, Makeig S. EEGLAB: An open source toolbox for analysis of single-trial EEG dynamics including independent component analysis.

Journal of Neuroscience Methods. 2004;**134**:9-21

[47] Vaughan CL. Are joint torques the Holy Grail of human gait analysis? Human Movement Science. 1996;**15**:423-443

[48] Hall M, Frank E, Holmes G, Pfahringer B, Reutemann P, Witten IH. The WEKA data mining software. ACM SIGKDD Explorations Newsletter. 2009;**11**:10-18

[49] Mannini A, Trojaniello D, Cereatti A, Sabatini A. A machine learning framework for gait classification using inertial sensors: Application to elderly, post-stroke and Huntington's disease patients. Sensors (Basel). 2016;**16**, **134**:1-14

[50] Ye Q, Xia Y, Yao Z. Classification of gait patterns in patients with neurodegenerative disease using adaptive neuro-fuzzy inference system. Computational and Mathematical Methods in Medicine. 2018;**2018**:9831252

Chapter 4

Multiscale Segmentation of Microscopic Images

Dimiter Prodanov

Abstract

The chapter introduces multiscale methods for image analysis and their applications to segmentation of microscopic images. Specifically, it presents mathematical morphology and linear scale-space theories as overarching signal processing frameworks without excessive mathematical formalization. The chapter introduces several differential invariants, which are computed from parametrized Gaussian kernels and their derivatives. The main application of this approach is to build a multidimensional multiscale feature space, which can be subsequently used to learn characteristic fingerprints of the objects of interests. More specialized applications, such as anisotropic diffusion and detection of blob-like and fiber-like structures, are introduced for two-dimensional images, and extensions to three-dimensional images are discussed. Presented approaches are generic and thus have broad applicability to time-varying signals and to two- and three-dimensional signals, such as microscopic images. The chapter is intended for biologists and computer scientists with a keen interest in the theoretical background of the employed techniques and is in part conceived as a tutorial.

Keywords: Laplacian of Gaussian, scale spaces, mathematical morphology, Fourier domain

1. Introduction

Neurophysiological data are very much variable, and while certain patterns are prominent and reproducible (e.g., action potentials, tissue textures, and cells) they by no means can be easily defined precisely in a quantitative way. Data are enriched with unwanted patterns having complicated temporal and spatial structure, which are misleadingly called "noise." Unlike the noise, natural objects have features on a limited number of spatial or temporal scales. This observation is the starting point of all available multiscale methods of analysis. The main focus of the chapter are digital images; however the presented approaches can be applied in the more simple setting of time-varying one-dimensional signals, such as voltage electrophysiological recordings. In the subsequent presentation, the images will always be considered as two-dimensional signals sampled on a rectangular spatial grid. The reason is that all common microscopic approaches acquire images on a plane of illumination; thus three-, four-, and five-dimensional images are essentially sets of correlated planar signals. The third dimension can represent depth, time, or an acquisition channel. Obviously, in the case of four and five dimensions, the number of combinations increases. Therefore, one cannot assume isotropic resolution of the transfer function for more than two dimensions. This situation introduces anisotropy in microscopic images, compared to other imaging modalities, such as magnetic resonance imaging or computer tomography. In microscopy, in such way, the default case is a planar image.

All physical signals are bounded and of finite duration. Such signals are acquired as discrete samples from an underlying physical process which, as an idealization, can be considered as continuous. The physical signals are naturally related to the properties of the measurement apparatus. As another idealization, these properties are described by a linear transfer function so that the measurement process becomes a convolution (denoted further by $\star$). On the other hand, the measurement is always contaminated by an unwanted signal, which is denoted broadly as "noise." The noise process can be identified with the nonlinearities of the measurement process. In many occasions because of its apparent irregularity in time, it can be treated as a purely random process. Since the physical measurement is a repeated process, the Gaussian noise comes as a very common and useful model by virtue of the *central limit theorem* of probability theory. The theorem roughly states that the weighted sum of uncorrelated random variables having finite variance approaches a normally distributed random variable. Hence, if the noise is not spatially and temporally correlated, methods suitable for treating Gaussian or Poisson noise are applicable. Evidently, very fast sampling or sampling from processes having long memories, such as viscoelastic interactions, can violate these requirements. In such settings, other noise models can become more suitable. The readers are directed to [1] for a useful noise classification.

The chapter is organized as follows. Section 2 discusses the segmentation problem in general. Section 3 gives an overview of the mathematical morphology theory and provides examples. Section 4 gives an overview of the geometrical image features from the perspective of differential geometry. Section 5 introduces several types of scale spaces and their application in segmentation. Section 6 discussed implementation details. The chapter is intended for biologists and data scientists with keen interest in theoretical background of the employed techniques and is in part conceived as a tutorial. The references cited in the chapter are suitable for introductions on the mentioned topics.

2. Brief overview of image segmentation approaches

Extraction of an object's boundaries from a digital image is called *segmentation*. Image segmentation is related also to object classification, which does not require precise delineation of the object boundary. Therefore, segmentation can be also viewed as classification on a pixel level.

The image segmentation is a nontrivial problem. For a successful image segmentation, it is important to have prior knowledge of the image composition, that is, the texture properties of the background and the objects of interest. Segmentation generates a mask consisting of a binary image delimiting the objects of interest present in the raw image. The challenge is to define an accurate segmentation methodology or at least an approach that enables segmentation of biologically relevant features. There are several classes of methods, which can be applied in different circumstances. These can be classified broadly into two classes: (i) intensity based, where the hypothesis is that only differences in the image intensity histogram can be sufficient for segmentation and (ii) geometry based, where the image is transformed so that the geometrical features of interest become enhanced.

Historically, the first and simplest segmentation methods are based on global thresholding of the histogram. Classical threshold-based methods consist of identifying a given pixel intensity level that allows for separating the object of interest from the background. There are about 40 different global thresholding algorithms [2]. Classically, an algorithm involving thresholding includes the following stages:

- Preprocessing steps, which decrease the spatial variation of the image
- Thresholding, which produces one or more binary masks
- Masking or region of interest (ROI) selection
- Post-processing steps, for example, including second thresholding or watershed

The watershed is based on a topographical interpretation of the grayscale image as terrain of mountains and valleys; the algorithm interpolates boundaries between objects based on the continuity in intensity peaks.

Various image filtering techniques can be introduced as preprocessing steps. These transformations are representations of mathematical operators. These operators have formal properties, which make them suitable for certain types of signals. For example, the convolution-based linear operators assume continuity of the sampled signal, while morphological operations do not. The notion of scale comes as the support of the sampled kernel, so multiscale analysis provides rules how the supports of different operators change with scale. It is also important to consider the sampling of the operators in the digital domain.

There are various geometry-based segmentation approaches, for example, using edges, distances, or texture statistics. In addition, there is a vast array of pre- and post-processing techniques, such as smoothing, mathematical morphology operations (i.e., watershed), partial differential equation methods, and shape methods. This manuscript will focus on the geometry-based methods with a particular emphasis on the edge detection techniques. Geometry-based approaches are invariant to changes of illumination, which is an issue in natural images and some microscopic techniques. In contrast, geometry-based approaches are susceptible to structural or texture "noise" so extra care must be taken to address such issues.

3. Mathematical morphology

The mathematical morphology (MM) theory is a way of analyzing objects' shapes by way of interaction with shape primitives called structuring elements (SE) or *kernels*. A structuring element can be thought of as a small window that scans the image and alters the central pixel within its frame. The mathematical morphology theory was developed by Matheron and Serra [3].

MM operators are useful for the analysis of both binary and grayscale images. Their common usages include edge detection, noise removal, image enhancement, and image segmentation. MM approaches employ topological transformations and hence do not depend, on the particular noise model. Therefore, they can be used also in situations, where the noise is non-Gaussian.

The main building blocks of MM are the *erosion* $\ominus$ and *dilation* $\oplus$ operators (see Appendix A.1). Erosion and dilation are best understood by their action on black-and-white images. If the white pixels represent the objects of interest, then after an erosion with a SE, the white objects are contracted as the SE is inscribed inside every white object. After a dilation with a SE, the white pixels are expanded as the SE is circumscribed outside every white object. The action on grayscale images is similar but must be understood in terms of ranking operations—that is, taking maxima and

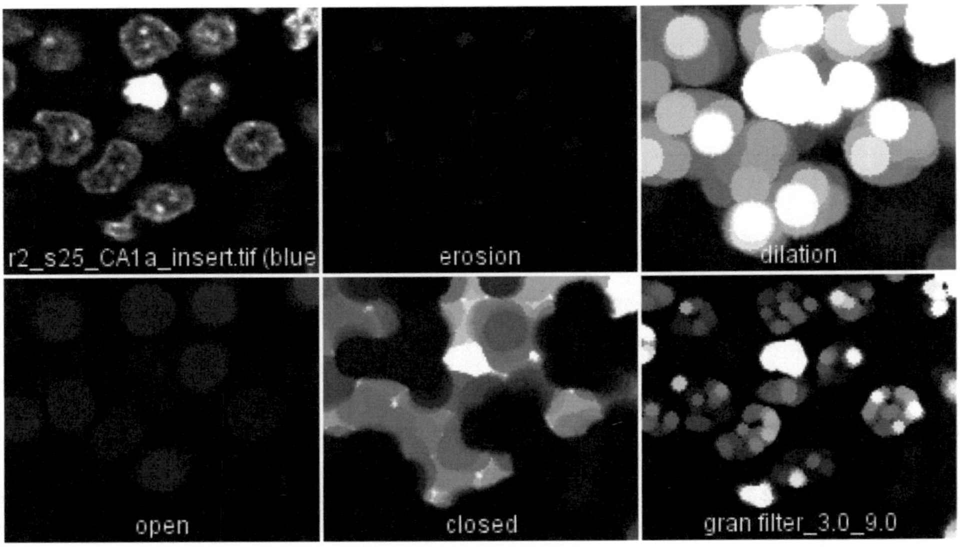

Figure 1.
Fundamental morphological operations. On the first row, an image of cell nuclei stained with DAPI (left) eroded (center) or dilated (right) with a disk of radius 10. On the second row, the same image is opened (left), closed (center), or granulometrically filtered. The inscribed numbers denote SE radii.

minima (**Figure 1**). The operations erosion and dilation can be composed into two more basic operations—opening and closing. The opening with a SE, denoted by E, is expressed as

$$I \circ E = (I \ominus E) \oplus E \tag{1}$$

The closing with a SE is expressed as

$$I \bullet E = (I \oplus E) \ominus E \tag{2}$$

The opening operation removes the objects, which are covered by E, while the closing, by duality, removes object's complement (i.e., holes in objects), which are covered by E. The so-developed theory is topological in nature because it does not depend explicitly on the concept of size but only on covering and inclusion. Classically, the MM theory was developed for uniform homothetic scaling of the SEs, but it can be extended to nonhomogeneous groups of scaling transformations. The scaling can be interpreted as generating a system of neighborhoods of every given point, thus reinforcing the topological interpretation. This gives rise to partial differential equation interpretation of the MM theory [4].

The multiscale aspects of the theory are due to the scaling of the structure elements. For example, the seed SE can be rescaled homothetically and then applied to the image. Such series of successive openings provides a measure of the prevalence of objects of a given size and is called granulometry (**Figure 2**). Granulometry can be used also to segment compact bright objects by means of a top-hat transform, where from the primitive image its opened version is subtracted:

$$T_E(I) = I - E \circ I, \quad G_{U,L}(I) = L \circ I - U \circ I \tag{3}$$

The second equation represents the granulometric filtering operation, which can extract bright objects of a specific size range from an image [5, 6].

Homogeneous scaling, that is, homothety, can be varied with the metric, which is induced on the SE. This can be box-like, circular, diamond, etc.

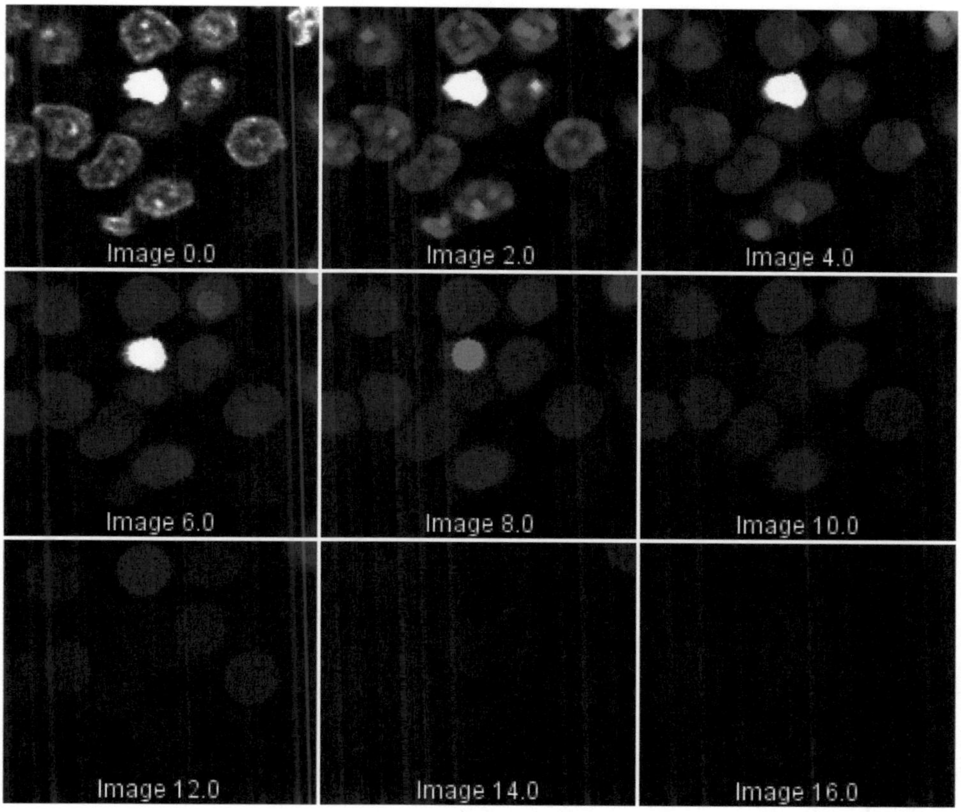

Figure 2.
Granulometry of cell nuclei. An image of cell nuclei stained with DAPI is opened with an increasing sequence of disk-shaped kernels. Note the eventual disappearance of the central bright object. The inscribed numbers denote SE radii.

Another useful realization is the morphological gradient operation, which is the difference between an opening and a closure $\delta_E(I) = I \circ E - I \bullet E$. It can be used to extract connected shapes by subsequent thresholding.

4. Geometrical image features

Mathematically, images can be represented as surfaces in the three-dimensional Euclidean space, where the elevation represents the signal intensity. In this sense, the intensity at a certain point in the direction $x + r$ can be interpolated from its local neighborhood as

$$I(x+r) = I(x) + r \cdot \nabla I + \frac{1}{2} r^T \cdot \mathbb{H} I \cdot r + \mathcal{O}\left(r^T r\right) \tag{4}$$

Components of the gradient are given by $\nabla u = \left(u_x, u_y\right)$. The Hessian tensor is given by the matrix

$$\mathbb{H} u = \begin{pmatrix} u_{xx} & u_{xy} \\ u_{yx} & u_{yy} \end{pmatrix} \tag{5}$$

where for smooth signals the partial derivatives commute $u_{xy} = u_{yx}$. This picture is a part of the so-called jet space—a higher dimensional differential descriptor space, as a natural basis for encoding the geometry of an image local neighborhood [7, 8]. The subscripted notation will be used to identify partial derivatives with respect to the coordinates.

The fact that digital images are sampled on a discrete grid may represent some difficulty as differentiation in the literal sense does not work for discrete signals. Notably, naive computations are numerically unstable and amplify the high-frequency noise. This difficulty can be overcome by applying the distribution theory, starting from the Leibniz identity for smooth signals [7]:

$$\nabla(I \star G) = (\nabla I) \star G + I \star \nabla G \tag{6}$$

where ∇ represents the gradient given by its principal components $\nabla = (\partial/\partial x, \ \partial/\partial y)$. For the whole space if the kernel vanishes fast at infinity, we have $(\nabla I) \star G = -I \star \nabla G$. Therefore, even for discrete images, by extension, one can define differentiation in terms of convolution with a differential of a kernel as

$$\nabla_G I := -I \star \nabla G \tag{7}$$

From this point on, differentiation of a digital image will be interpreted only in the generalized sense as a convolution with some smooth kernel. In such way, various local differential geometric invariants can be also incorporated into the processing. There are several filter families, which possess desirable properties, which can be exploited for systematic image noise suppression and computation of differential invariants. These families are formalized by the framework of the scale-space theory. Notable examples are the spatial derivatives of the Gaussian, which are used in the linear scale-space theory 5.1.

4.1 Differential invariants

There are several types of geometric features, which are useful for segmentation applications. Typical interesting image features are blobs, filaments, and corners. Notably, object boundaries can be represented in terms of edges, which can be approximated by steps in image intensity. All these features can be computed from the local differential structure of the image. The theory will be exemplified with the Gaussian derivatives, which, in view of the duality property of Eq. (7), can be used to compute the image derivatives.

The first four differential invariants are given in **Table 1**. The gradient vector field of the test image is represented in **Figure 3**.

The eigenvalues of the Hessian tensor are solutions of the characteristic equation $\det(\mathbb{H} - \lambda \mathbb{I}) = 0$, where $\mathbb{I}$ is the identity matrix. This is a square equation with two real roots $\lambda_{1,2}$, such that $\lambda_1 + \lambda_2 = \Delta_G$ and $\lambda_1 \lambda_2 = \det \mathbb{H}$. If both eigenvalues are

Gradient amplitude	$A = \sqrt{G_x^2 + G_y^2}$
Gradient orientation	$\sin\phi = G_y / \sqrt{G_x^2 + G_y^2}$
	$\cos\phi = G_x / \sqrt{G_x^2 + G_y^2}$
Laplacian	$\Delta_G = Tr\,\mathbf{H} = G_{xx} + G_{yy}$
Determinant of the Hessian	$\det \mathbf{H} = G_{xx} G_{yy} - G_{xy}^2$

Table 1.
Second-order differential invariants.

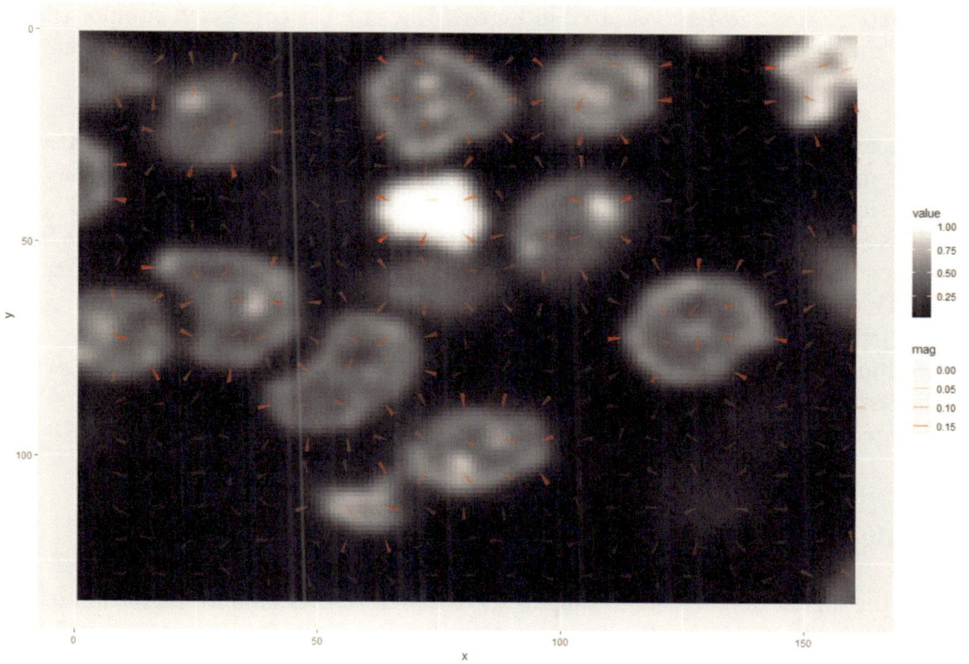

Figure 3.
The gradient image field. The gradient vector filed is overlaid onto a smoothed and downsampled version of the original image. The gradient amplitude is encoded by the arrow intensity.

Figure 4.
Connected components of the Laplacian operator's zero space. The boundary (left) is overlaid on the cell nuclei image (right). The connected components (center) are calculated from Laplacian of Gaussian, s = 12.

negative, this is an indication for a bright blob-like feature around the point of reference. In a similar way, if both eigenvalues are positive, there is a dark blob-like feature around the point of reference.

If the eigenvalues have opposite signs, this is an indication for a saddle point at the point of reference. Therefore, the zero-crossing of the Laplacian operator can be used to delimit regions, encompassing blobs. The zero-crossings form the so-called zero space, which can be used to identify objects. The regions where the Laplacian changes sign can be extracted by connected component analysis, which are defined as regions of adjacent pixels that have the same input label. In this regard, different neighborhoods can be considered for the blobs (4-connected, N4) and for the contours (8-connected, N8). To compute the connected components of an image, we first (conceptually) split the image into horizontal runs of adjacent pixels and then color the runs with unique labels, reusing the labels of vertically adjacent runs whenever possible. In a second phase, adjacent runs of different colors are then merged [9].

The zero space is demonstrated in **Figure 4**, where the connected components where the Laplacian changes sign are labeled. From the figure, it is apparent that the cell nuclei can be enclosed well by the blobs.

The number of differential invariants increases with the increase of the image dimensions. However, the theory can be extended along similar lines. A very useful development in this direction is geometric algebra and calculus, which provide a dimension invariant representation of the geometrical structures.

The so-introduced geometric image features can be used as building blocks for advanced machine learning strategies for interactive segmentation and classification. This strategy was implemented in two segmentation platforms based on ImageJ/Fiji. The Trainable Weka Segmentation (TWS) [10] and the Active Segmentation [11] have recently presented new opportunities for analyzing complex datasets. Specifically, the active segmentation uses the scale-space-based filters presented here.

5. Scale-space theory

In the digital domain, smoothing leads to loss of resolution and, therefore, of some information. However, the information loss can be limited if one uses multiple smoothing scales (see **Figure 5**).

Scale-space theory is a framework for multiscale image representation, which has been developed by the computer vision scientists with intuitive motivations

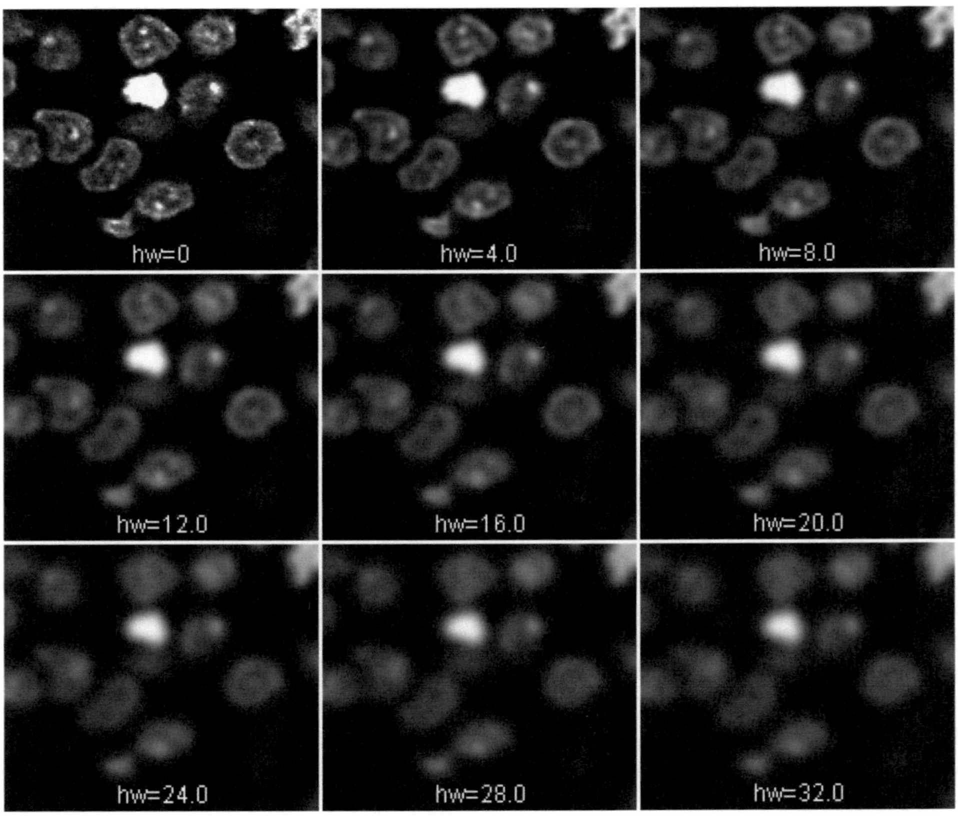

Figure 5.
Gaussian scale space of cell nuclei. An image of cell nuclei stained with DAPI is convolved with an increasing sequence of Gaussian kernels. Inscribed labels denote kernel half widths.

from physics and biological vision, as introduced by [12]. The underlying idea is to account for the multiscale nature of real-world objects, which implies that objects may be perceived in different ways depending on the scale of observation. Taken to the limit, a scale-space representation furthermore considers representations at all scales simultaneously. Scale spaces have been introduced independently in Japan and Europe by [12, 13]. The axiomatic linear scale-space theory was formalized in series of works by Witkin [14] and Koenderink [15].

Scale-space approaches are ubiquitous in feature detection/description, as well as dense correspondence mapping (e.g., large-offset optical flow is typically done in coarse-to-fine fashion) [9].

5.1 The Gaussian scale space

The liner scale-space theory provides a systematic way of dealing with spatially uncorrelated Gaussian noise. A fundamental result of scale-space theory states that if some general conditions are imposed on the types of computations that are to be performed in the earliest stages of visual processing, then convolution by the Gaussian kernel and its derivatives provides a canonical class of image operators with unique properties. The Gaussian kernel in 1D is given by

$$G(x) = \frac{1}{\sqrt{2\pi s}} e^{-\frac{x^2}{2s}} \tag{8}$$

and

$$G(x, y) = G(x)G(y) = \frac{1}{2\pi s} e^{-\frac{x^2+y^2}{2s}} \tag{9}$$

in two dimensions. A very useful property of the kernel is its separability, which allows for efficient computation of convolutions for multiple spatial dimensions. That is, for example, in two dimensions

$$G(x, y) \star I = G(x) \star (G(y) \star I) = G(y) \star (G(x) \star I) \tag{10}$$

Therefore, the computational cost scales linearly with the support of the kernel rather than quadratically.

The Gaussian scale space depends on a free scalar parameter s representing the scale of possible structures in the image [12–15]. In the typical implementation of the theory, the scale parameter enumerates a space of smooth Gaussian test kernels of rapid decay, which are convolved with the digital image. In one dimension, Gaussian smoothing implies that new local extrema or new zero-crossings cannot be created with increasing scales. Gaussian kernels provide several advantages: (i) they are rotationally invariant, (ii) they do not produce artificial extrema in the resulting image, and (iii) successive convolutions with different kernels can be combined. Mathematically, this imposes a very useful semigroup structure, equivalent to the heat/diffusion equation. In this sense, the image structures diffuse or "melt down," so that the rate of this diffusion indicates the "robustness" of the structure.

In its typical presentation, the scale-space theory applies only smoothing steps. Later, the theory was extended to include also differentiation and thus account for the differential structure of the images [16]. In the spatial domain, the Gaussian derivatives for the one-dimensional case can be computed in closed form as

$$G_n(x) = \frac{\partial^n}{\partial x^n} G(x) = \frac{(-1)^n}{\sqrt{2\pi s^{n+1}}} He_n\left(x/\sqrt{s}\right) e^{-\frac{x^2}{2s}} \tag{11}$$

where $He_n(x)$ is the statistician's Hermite polynomial of order n. The sequence of statistician's Hermite polynomials satisfies the recursion

$$He_{n+1}(x) = xHe_n(x) - nHe_{n-1}(x) \tag{12}$$

starting from $He_0(x) = 1$ and $He_1(x) = x$. This allows for efficient simultaneous computation of all derivatives up to an order n in order to populate the n-jet space. The n-jet components can be used to build the differential invariants up to order n. An example is presented in **Figure 6**, where the five unique components of the Gaussian jet-2 space are computed. The original dataset is present in the ImageJ public image database.

In spite of several properties that make linear diffusion filtering useful, it also reveals some drawbacks [17]:

1. An obvious disadvantage of Gaussian smoothing is the fact that it does not only smooth noise but also blurs important features such as edges. Moreover, it is uncommitted to any prior information about the image structure.

2. Linear diffusion filtering propagates edges when moving from finer to coarser scales, which can lead to difficulties in edge identification and instabilities.

5.2 α-Scale spaces

The α-scale spaces introduce nonlinearity on the level of differentiation. Notably, the Gaussian differentiation is replaced by another convolution operation, involving a power law. Pauwels et al. [18] and later Duits et al. [19] investigated the use of fractional powers of the Laplacian in connection with scale invariant smoothing and scale-space theory, respectively. This approach tries to overcome some of the limitations of the Gaussian scale spaces identified above. The evolution

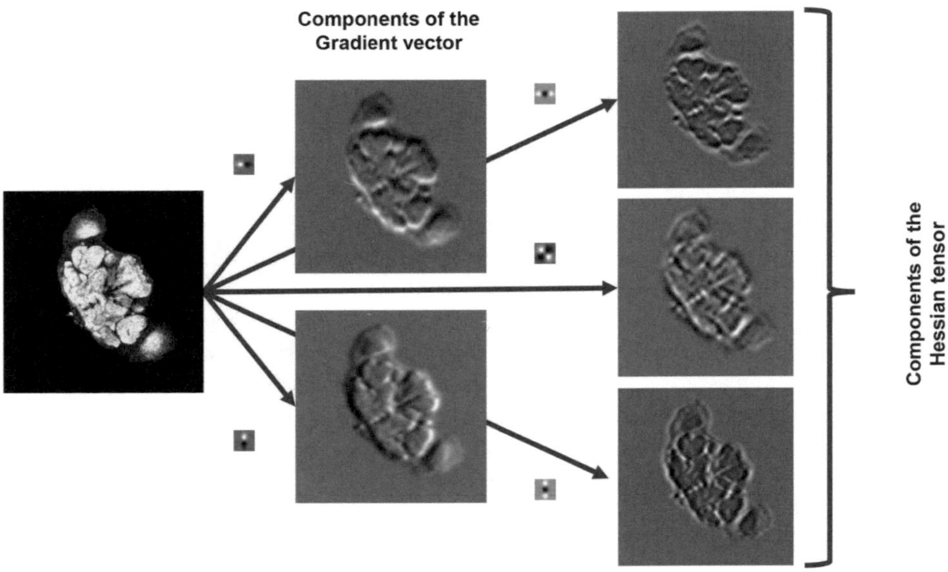

Figure 6.
Differential Gaussian 2-jet space. A microscopic image of Drosophila brain (first column) is convolved with Gaussian derivative kernels. Different kernels are shown above the arrows. The second column shows the components of the gradient. The third column shows the components of the Hessian. The local jet space of order k has $k(k+1)/2$ different components.

is governed by two parameters—the scale s and the order of differentiation α. The approach leads to formulation and solving of a fractional heat problem:

$$\begin{aligned} u(0, \mathbf{x}) &= I(\mathbf{x}) \\ u_s(s, \mathbf{x}) &= -(-\Delta)^{\alpha/2} u(s, \mathbf{x}), \quad 1 \le \alpha \le 2 \end{aligned} \tag{13}$$

The Riesz fractional Laplacian operator is defined in the Fourier domain by

$$(-\Delta)^{\alpha} U(\mathbf{k}) := |k|^{\alpha} U(\mathbf{k}) \tag{14}$$

where the $|k| = \sqrt{\mathbf{k} \cdot \mathbf{k}}$ is the modulus of the wave vector $\mathbf{k}$. In this way, the solution can be expressed in terms of a convolution with a very general special function—the Wright function [20]. Numerical routines for computation of the Wright function are still not readily available; therefore the computations is easier achieved using fast Fourier transform (FFT) and its inverse, IFFT.

5.3 Nonlinear scale spaces

Linear diffusion scale spaces are well-posed and have a solid axiomatic foundation. On the other hand, for some applications, they have the undesirable property that they do not permit contrast enhancement and that they may blur and delocalize structures. Nonlinear scale spaces try to overcome some of these limitations. Such scale spaces arise in nonlinear partial differential equation framework, which will be sketched below. The formal properties of some types of scale spaces have been established by Alvarez et al. [4]. In particular, they established a strong link with the related field of mathematical morphology (see Section 3). The following second-order partial differential equation was demonstrated in particular

$$u_t = F(\mathbb{H}u, \nabla u), \quad u(0, x) = f(x) \tag{15}$$

where Hu are the components of the Hessian tensor, ∇u represents the components of the gradient, and $f(x)$ is the original image. It is interesting that MM operations can also be represented in this framework as $u_t = \pm \|\nabla u\|$ for dilation and erosion, respectively.

In this line of development, the Laplacian of Gaussian (LoG) operator can be decomposed into orthogonal and tangential components ([17], Ch. 1). The representation is provided below:

$$\begin{aligned} \Delta_G &= G_{xy} + G_{yy} = \Delta_{|G} + \Delta_{\perp G} \\ \left(G_x^2 + G_y^2\right)\Delta_{\perp G} &= (G_x^2)G_{xx} + (2G_xG_y)G_{xy} + \left(G_y^2\right)G_{yy} \\ \left(G_x^2 + G_y^2\right)\Delta_{\|G} &= (G_x^2)G_{xx} - (2G_xG_y)G_{xy} + \left(G_y^2\right)G_{yy} \end{aligned} \tag{16}$$

The parentheses denote scalar multiplication with the component of the gradient. The orthogonal decomposition is equivalent to an effective vectorization of the filter. The normal component is antiparallel to the gradient (i.e., in normal direction to the isophote curve), while the tangential component is parallel to the isophote curve passing through the point. These components can be used to segment blob-like or tubular structures. Segmentation based on the orthogonal decomposition is illustrated in **Figure 7**.

The orthogonal decomposition leads naturally to anisotropic diffusion (**Figure 8**). For example, if the tangential component is selected, this will lead to

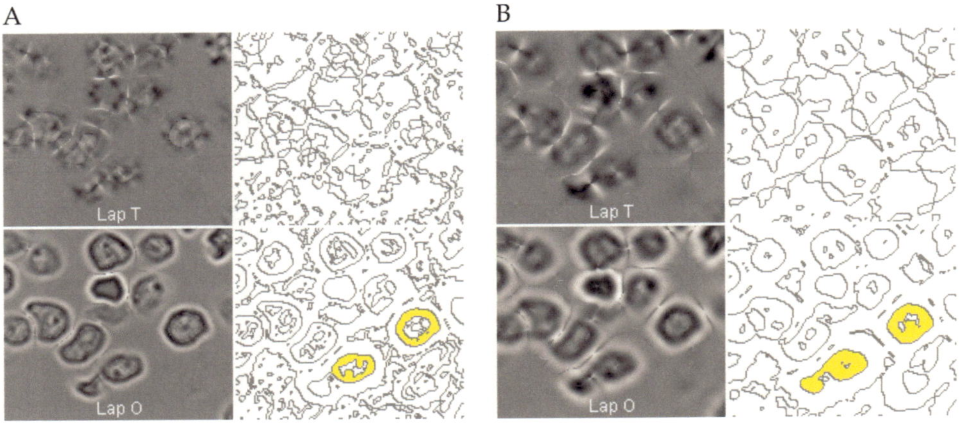

Figure 7.
Blob segmentation. Zero-crossing of the LoG decomposition, s = 6 (A) and s = 12 (B). Two blobs are highlighted for better appreciation. The normal component is in Lap T, tangential component of the Laplacian; Lap O, normal component of the Laplacian.

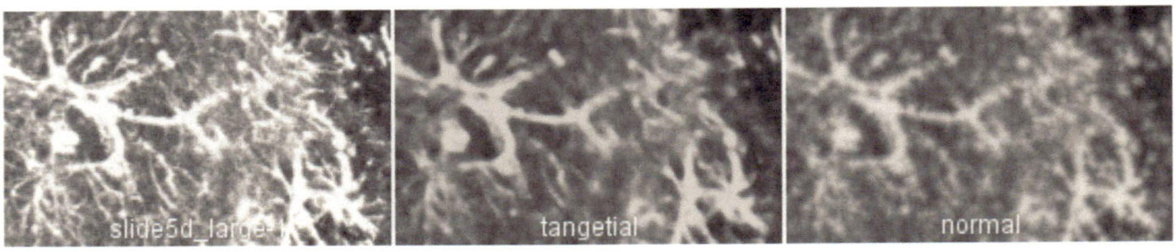

Figure 8.
Anisotropic diffusion along principal flow directions. Astrocytes were stained immunohistochemically for glial fibrillary acidic protein (GFAP) and imaged on a confocal microscope (left). Anisotropic diffusion evolved according to the orthogonal decomposition of the Laplacian, s = 3, 3 steps—Tangential direction (center) and along the gradient direction (right). Note the granularity of the right image and its blurred appearance compared to the central image.

preservation of globular structures, while if the normal component is selected, this will lead to enhancement of the tubular structures. The equation $u_t = \Delta_{\perp G} u$ is called *mean curvature motion* equation [17]. An example is presented in **Figure 8**.

6. Implementation

The filters described in the present manuscript are implemented in ImageJ as a set of plug-ins (**Table 2**). Two implementation strategies have been used: the integer order filters are implemented in the spatial domain, while the fractional order filters are implemented in the Fourier domain [21].[1] The plug-ins are distributed under the GNU Lesser General Public License v3.0 and are available as code repository from the GitHub website [22].[2] The choice of implementation platform was due to the widespread use of ImageJ in the biomedical and life science communities.

ImageJ is a public domain image processing program written in Java. Since its inception in 1997, ImageJ has evolved to become a standard analytical tool in a

---

[1] The installation procedure of the spatial-domain filters is straightforward, and this is the reason why only spatial-domain filters are included in the public repository.

[2] https://github.com/dprodanov/scalespace.

Plug-in	Function
LoG filter	Laplacian of Gaussian (LoG)
ALoG filter	Anisotropic decomposition of LoG
ADiff filter	Anisotropic diffusion
LoG2 filter	Bi-Laplacian of Gaussian
LoGN2D filter	N-order power of the Laplacian of Gaussian
Gaussian jet	Gaussian jet of order n
Zero-crosser	Connected components

Table 2.
ImageJ plug-ins demonstrated in the chapter.

number of scientific communities. In particular, for life science communities, it is available as the Fiji plug-in platform, which allows for easy plug-in deployment and dependency management. ImageJ has an open architecture providing extensibility via third-party Java modules (called plug-ins) and scripting macros. It is developed by Wayne Rasband since 1997 and expanded via contributed software code by an international group of contributors. Plug-ins are distributed together with their source code under various licenses determined by the plug-in authors. The user guide of the platform [23] is maintained at http://imagej.nih.gov/ij/docs/guide. Public resources are available on the ImageJ website and the ImageJ Information and Documentation Portal https://imagejdocu.list.lu/. In addition, textbook introductions to image processing with ImageJ can be found in [24].

7. Discussion

The morphological complexity of the nervous tissue is a challenge for conventional segmentation techniques developed for computer vision applications or cultured cells. The challenges lie in the morphological complexity of neurons and glial cells overlaid on the heterogeneity of the extracellular matrix. This complexity translates into variations of the tracer signal and touching of relevant structures.

Segmentation of fluorescent images poses particular issues due to low signal-to-noise ratio, unequal staining, as well as the complexity of structures that need to be identified. This irreducible variation must inform choices about segmentation methods. In particular, methods employing multiple spatial scales are favorable. Structure identification is inherently a multiscale problem because object structure is recursive, that is, objects may contain substructures, which themselves contain substructures, etc.

A large number of algorithms for image segmentation have been proposed in literature (overview in [9]). However, many of them completely ignore the issue of scale. As a result, they are capable of identifying only limited types of structures. In contrast, multiscale approaches eventually rely on the topological properties of the segmented objects, either by means of scale spaces or by nonlinear vector field transforms [25, 26]. As a result, such methods are able to combine detected features into robust segmentation tools. The present chapter introduced two classes of multiscale methods for image segmentation: the mathematical morphology operations and scale spaces. The main applications of the theory are classification and segmentation of signals. Presented methods are generic and thus have broad applicability to both one-dimensional signals, such as electrophysiological recordings, and to two and three-dimensional signals, such as microscopic images.

8. Conclusions and outlook

The main utility of the presented approaches is to build a multidimensional multiscale feature space, which is subsequently used to learn characteristic "fingerprints" of the objects of interests. The large variation of structures present in microscopic images precludes the design of an "ideal" tool. Instead, multiple approaches should be combined and features computed that would inform machine learning approaches, which are able to adapt to the morphology of the cells and tissues at hand. Development in this direction has been undertaken with the advent of deep learning techniques. ImageJ-based implementations, such as the Trainable Weka Segmentation [10] and the Active Segmentation platforms [11], have been made available to end-users.

Acknowledgements

The author declares no conflict of interest.

List of acronyms

MM	mathematical morphology
SE	structuring element
FFT	fast Fourier transform
IFFT	inverse fast Fourier transform
GFAP	glial fibrillary acidic protein
LoG	Laplacian of Gaussian
ROI	region of interest

A. Appendix

A.1 Ranking operations

This section starts with a brief introduction to the set notation. In many sources it is called also the "set builder notation." The empty set is denoted as ∅. A set containing only one member (singleton, for example the number 7) is denoted as {7}. A set consisting of members fulfilling certain condition (in the sense of a predicate function) is denoted as $X = \{x : \text{predicate}(x)\}$. For example, all positive reals smaller than 7 are denoted as $X = \{x : x > 0, x < 7\}$.

From a formal perspective, the mathematical morphology is the application of lattice theory to spatial structures [3]. Formally, the erosion is expressed as

$$I \ominus E = \{x : x + b \in I, b \in E\} \tag{17}$$

for binary images, while for grayscale discrete images, it is

$$I \ominus E = \min_{y \in B} (I(x + y) - B(y)) \tag{18}$$

Formally, the dilation for binary images is

$$I \oplus E = \{x : x - b \in I, b \in E\} \tag{19}$$

while for grayscale discrete images

$$I \oplus E = \max_{y \in B} (I(x+y) + B(y)) \tag{20}$$

In the continuous approximation case, the minimum and maximum should be replaced by infimum and supremum, respectively. Consider the set X. The *infimum* inf, or the *greatest lower bound*, is the greatest number, which is not necessarily in X but is smaller (or equal) than all members of X. For a finite set, the *infimum* coincides with the minimum. The *supremum* sup, or the *least upper bound*, is the smallest number, which is not necessarily in X but is greater (or equal) than all members of X. For a finite set, the *supremum* coincides with the maximum.

A.2 Some useful Fourier transforms

The concept of frequency and the decomposition of waveforms into elementary "harmonic" wave motions first arose in the context of music and sound. The Fourier transform and its inverse in the continuous domain are defined as

$$U(\mathbf{k}) = \mathcal{F}u(\mathbf{x}) = \iint_{-\infty} e^{-2\pi i k \cdot x} u(x) dx^2 \Leftrightarrow u(\mathbf{x}) = \mathcal{F}^{-1}U(\mathbf{k}) = \iint_{-\infty} e^{2\pi i k \cdot x} U(k) dk^2 \tag{21}$$

The reader is directed to the book of [27] for an introduction on the topic. The Fourier transform of the Gaussian is given by

$$\tilde{G}(f) = e^{-2\pi^2 s f^2} \tag{22}$$

and of its derivatives by

$$\tilde{G}_n(f) = (i2\pi f s)^n e^{-2\pi^2 s f^2} \tag{23}$$

In the Fourier domain, the fractional heat kernel is expressed as

$$\tilde{G}(s, \omega, \eta) = e^{-(2\pi)^\alpha |k|^{2\alpha} s} \tag{24}$$

Integer-order powers. Integer powers of the Laplacian operators are successive compositions of the Laplacian operator [28]:

$$2^n \tilde{L}_n(\omega, \eta) = (-1)^n (2\pi s)^{2n} e^{-2\pi^2 s k^2} k^{2n} \tag{25}$$

By substitution with the radial wavenumber $k = \sqrt{\omega^2 + \eta^2}$, it can be demonstrated that the kernel is radially symmetric about the vector k and gets sharper with increasing the order n:

$$\tilde{L}_n(k) = (-1)^n 2^n \pi^{2n} k^{2n} e^{-2\pi^2 k^2 s} \tag{26}$$

Fractional-order powers. In the fractional domain the operator can be expressed as a direct generalization:

$$\tilde{L}_{n,\alpha}(k) = (-1)^n 2^n \pi^{2n} k^{2\alpha} e^{-2\pi^2 |k|^{2\alpha} s} \tag{27}$$

Therefore, the kernel bandwidth can be controlled by the fractional power α.

A.3 Convolutions and Fourier domain processing

So far the described theory can work both in the spatial and the Fourier domain. A schematic treatment of the Fourier transform is given in Section A.2. Interested readers are referred to Burgers for a more complete introduction. The Fourier domain processing implemented via Fast Fourier Transform has a certain advantage. On the first place, for large convolution kernels, it can lead to speedup. This is so because convolution in spatial (respectively temporal) domain corresponds to multiplication in the Fourier domain. This incurs fixed computation costs; therefore, the convolution operation scales as $N\log(N)$, where N is the size of memory occupied by the digital image. Therefore, the following processing scheme becomes useful:

$$\begin{aligned} \text{FFT}: &\quad I \mapsto I_F \rightarrow \underbrace{K_F \cdot F_I}_{J_F} \\ \text{IFFT}: &\quad J_F \mapsto J \leftarrow I * K \end{aligned} \tag{28}$$

In the diagram above, the arrows indicate transformation, while FFT and IFFT denote forward and Inverse Fast Fourier Transforms, respectively. In the example of differentiation in the previous section, the kernel is the wave vector $K_F = \mathbf{k} = (k_x, k_y, k_z)$.

Author details

Dimiter Prodanov[1,2]

1 Environment Health and Safety, Imec, Leuven, Belgium

2 Neuroscience Research Flanders, Leuven, Belgium

*Address all correspondence to: dimiterpp@gmail.com; dimiter.prodanov@imec.be

References

[1] Boyat AK, Joshi BK. A review paper: Noise models in digital image processing. Signal & Image Processing: An International Journal. 2015;**6**(2): 63-75

[2] Sezgin M, Sankur B. Survey over image thresholding techniques and quantitative performance evaluation. Journal of Electronic Imaging. 2004; **13**(1):146-166

[3] Serra J. Introduction to mathematical morphology. Computer Vision, Graphics, and Image Processing. 1986; **35**(3):283-305

[4] Alvarez L, Guichard F, Lions PL, Morel JM. Axioms and fundamental equations of image processing. Archive for Rational Mechanics and Analysis. 1993;**123**(3):199-257

[5] Prodanov D, Heeroma J, Marani E. Automatic morphometry of synaptic boutons of cultured cells using granulometric analysis of digital images. Journal of Neuroscience Methods. 2006; **151**:168-177

[6] Prodanov D, Feirabend HKP. Automated characterization of nerve fibers labeled fluorescently: Determination of size, class and spatial distribution. Brain Research. 2008;**1233**: 35-50

[7] Florack L, Ter Haar Romeny B, Viergever M, Koenderink J. The Gaussian scale-space paradigm and the multiscale local jet. International Journal of Computer Vision. apr 1996;**18**(1): 61-75

[8] Larsen ABL, Darkner S, Dahl AL, Pedersen KS. Jet-based local image descriptors. In: Computer Vision – ECCV. Springer Berlin Heidelberg; 2012. pp. 638, 2012-650

[9] Szeliski R. Computer Vision. Springer-Verlag GmbH; 2010

[10] Arganda-Carreras I, Kaynig V, Rueden C, Eliceiri KW, Schindelin J, Cardona A, et al. Trainable weka segmentation: A machine learning tool for microscopy pixel classification. Bioinformatics. 2017;**33**(15):2424-2426

[11] Vohra S, Prodanov D. Classification and segmentation of cells in anatomic & time lapse microscopic images based on geometrical features and machine learning. Frontiers in Neuroinformatics. 2016;**10**

[12] Marr D, Hildreth E. Theory of edge detection. Proceedings of the Royal Society of London. Series B. 1980;**207**: 187-217

[13] Iijima T. Theory of pattern recognition. Electronics and Communications in Japan. 1963:123-134

[14] Witkin AP. Scale-space filtering. In: Proceedings of the 8th International Joint Conference on Artificial Intelligence (IJCAI '83), Vol. 2, Aug. 8–12. 1983. pp. 1019-1022

[15] Koenderink JJ. The structure of images. Biological Cybernetics. 1984; **50**(5):363-370

[16] Lindeberg T. Scale-Space. Wiley Encyclopedia of Computer Science and Engineering; 2007. pp. 2495-2504

[17] Weickert J. Anisotropic Diffusion in Image Processing. ECMI Series. Stuttgart: Teubner-Verlag; 1998

[18] Pauwels EJ, Van Gool LJ, Fiddelaers P, Moons T. An extended class of scale-invariant and recursive scale space filters. IEEE Transactions on Pattern Analysis and Machine Intelligence. 1995;**17**(7):691-701

[19] Duits R, Felsberg M, Florack L, Platel B. α-Scale spaces on a bounded

domain. In: Scale Space Methods in Computer Vision. Springer; 2003. pp. 494-510

[20] Mainardi F. Fractional Calculus and Waves in Linear Viscoelasticity. Imperial College Press; 2010

[21] García-Fernández V, Prodanov D. Image segmentation based on frequency domain operation (FFT) and plugin implementation for ImageJ. Frontiers in Neuroinformatics. 2015;**9**

[22] Prodanov D. Scale Space Analysis Filters for ImageJ. July 2019

[23] Ferreira T, Raspband W. ImageJ User Guide. NIH; 2012

[24] Pascau J, Mateos JM. Image Processing with ImageJ. PACKT PUB; 2013

[25] Ahuja N. A transform for multiscale image segmentation by integrated edge and region detection. IEEE Transactions on Pattern Analysis and Machine Intelligence. 1996;**18**(12):1211-1235

[26] Tabb M, Ahuja N. Multiscale image segmentation by integrated edge and region detection. IEEE Transactions on Image Processing. 1997;**6**(5):642-655

[27] Burger W, Burge M. Principles of Digital Image Processing. Springer London Ltd; 2009

[28] Prodanov D, Konopczynski T, Trojnar M. Selected applications of scale spaces in microscopic image analysis. Cybernetics and Information Technologies. 2015;**15**(7):5-12

Chapter 5

Empirical Mode Decomposition of EEG Signals for the Effectual Classification of Seizures

Fasil OK and Reghunadhan Rajesh

Abstract

Empirical mode decomposition (EMD) is a remarkable method for the analysis of nonlinear and non-stationary data. EMD will breakdown the given signal into intrinsic mode functions (IMFs), which can represent natural signals effectively. In this work, the competence of EMD with traditional features to classify the seizure and non-seizure EEG signals is studied. Due to the complex nature of human brain, the EEG signals which are recorded from different regions of brain are non-stationary in nature. Different features such as entropy features (approximate entropy (ApEn), sample entropy (SmEn), Shannon entropy (ShEn), Rényi entropy (RnEn)), fractal dimension features (Petrosian fractal dimension, Higuchi fractal dimension, Katz fractal dimension), statistical features (mean, standard deviation and energy) and exponential energy features are extracted from IMFs and fed to a SVM classifier. The performances of extracted features are studied independently. The result shows that, the EMD method is well suited for complex seizure EEG signal classification.

Keywords: seizures, EEG, empirical mode decomposition, intrinsic mode functions

1. Introduction

Seizures are characterized as unexpected, unprovoked and uncontrolled explosion of electrical impulses in brain [1]. During the seizure, the patient may experiences changes in behavior, loss of consciousness, unusual movements and unusual feelings [2, 3]. The recurrent and unprovoked seizure leads to epilepsy disorder which is a prevalent neurological disorder. Epilepsy disorder will tamper the patients way of life with social stigma, work productivity lose and premature death [4].

Electroencephalogram (EEG) is one of the traditional and easiest tool for the identification and diagnosis of seizures [5]. The availability of EEG for common people within their budgetary limits made it a typical method. Due to the sophisticated nature of brain system, the EEG signals acquired from the brain are also complicated. Automated analysis of EEG signals using modern signal processing techniques might be effortless and precise for the diagnosis of seizures rather than manual approach [6].

Out of modern signal processing techniques, empirical mode decomposition (EMD) is one of the widely used techniques for the efficient interpretation of signals and images. After the introduction of EMD by Huang [7] in 1998, several studies utilized the EMD for various applications. In Nunes et al. [8] used the EMD for texture analysis and image filtering. They have used bi-dimensional EMD in their method. In another work Zeng et al. [9] applied EMD for the effective classification of gait patterns between patients with Parkinson disease and healthy subjects. In another work Hasan et al. [10] combined the deep learning methods with EMD to classify cardiovascular disease. Xiwei et al. [11] utilized the advantages of EMD in a wind speed prediction model, in which, authors used EMD for the extraction of fluctuation features of wind speed data. Another important study by Thilagaraj et al. [12] also used EMD for the identification of alcoholism.

The usefulness of the empirical mode composition for the effective understanding of the EEG signal is proven in many works in the literature. In [13], authors classified the level of autism severity from EEG with the help of EMD. They have used artificial neural network for the classification of extracted feature from intrinsic mode functions (IMFs). Two-class motor imagery EEG signals are classified in another important study based on EMD [14]. Similarly Gaur et al. [15] used multivariate empirical mode decomposition for the effective classification of multi-class BCI by analyzing EEG signals.

In this work, we have studied the effectiveness of empirical mode decomposition for the classification of seizures by analyzing EEG signals. The filtered EEG signals are segmented into 10 non-overlapping segments and decomposed into IMFs using EMD. First four IMFs are used for the feature extraction. Various features such as approximate entropy, sample entropy, Shannon entropy, Rényi entropy, exponential energy, fractal dimensional features and statistical features (mean, standard deviation and energy) are extracted from the IMFs. Support vector machine (SVM) with RBF kernel is used for classifying the seizure.

The remaining sections of the paper are as follows. A short description of EMD and algorithm is explained in Section 2. Section 3 explains the details of the dataset used in this study and in Section 4; various feature extraction methods are mentioned. In Section 5 experimental setup and results are explained. A detailed discussion of achieved results is given in Section 6 and Section 7 concludes the paper.

2. Empirical mode decomposition (EMD)

Empirical mode decomposition is a data-driven decomposition method proposed by Huang et al. for the analysis of nonlinear and non-stationary data [7], which will decomposes the signal into finite and smaller number of intrinsic mode functions (IMFs). A non-stationary signal can be represented as sum of IMFs and each IMFs should follow two conditions: (1) the number of extrema and number of zero crossing of the IMFs should be equal or differ at most by one and (2) the mean value of two envelopes defined by local maxima and local minima should be zero [16].

IMFs can be extracted from a signal through a iterative method known as shifting process as follows:

1. Use cubic spline interpolation method to construct upper (e_{max}) and lower (e_{min}) envelops by connecting detected maxima and minima individually from the signal $x(t)$.

2. Calculate the mean $m(t) = \frac{(e_{\max}+e_{\min})}{2}$.

3. Extract the difference $d(t)$ between signal $x(t)$ and calculated $\mathrm{m_1(t)}$, $d(t) = x(t) - m(t)$.

4. Check whether $d(t)$ is satisfying the IMFs basic conditions. Repeat step 1 to 3 until $d(t)$ satisfying the IMFs conditions.

5. if $d(t)$ satisfies IMFs condition once, define the first IMF as $IMF_1 = d(t)$.

6. The next IMFs can be obtained by generating residue $r(t)$ as $r(t) = x(t) - IMF_1$ and use these residue as the original data for the next iteration.

7. Iteration will stop when final residue is a function which cannot produce any more IMFs or final residue is constant/monotonic function.

The original signal can be represented as the sum of all IMFs and final residual.

$$x(t) = \sum_{i=1}^{K} IMF_i + r_K(t) \tag{1}$$

where IMF_i is the ith IMF, K is the number of IMF and $r_K(t)$ is the final residual.

3. Dataset

A benchmark data set named as Bern-Barcelona EEG dataset is used in this study. The dataset includes two class EEG signals such as focal and non-focal. Each class contains 3750 pairs of signals. EEG signals in the focal class are collected from the epileptic area of the brain and non-focal signals are collected from non-epileptic area of the brain. The signals are 20 s duration with 10,240 samples in each. The signals are sampled at 512 Hz sampling rate. In our study we have used 50 signals from each class as did in many other studies [17–19].

4. Feature extraction

Feature extraction is one of the important tasks in any machine learning application. An effective and unbiased feature will provide the best results. There are several features, which are traditionally used for various EEG related studies.

Entropy features are widely used for the analysis of various non-stationary biosignals [20–22]. Different verities of entropy are introduced in past years. In this work we have used four verities of entropy features, namely approximate entropy (ApEn), sample entropy (SmEn), Shannon entropy (ShEn) and Rényi entropy (RnEn). Among considered entropy features, approximate entropy introduced by Pincus [23] is a good measure of complexity for non-stationary signals. One of the study proposed by Hozinger et al. [24], extracted approximate entropy from ECG time-series for better understanding of electrocardiogram (ECG) signals. Another study by Ahmed et al. [25] utilized approximate entropy for surface electromyogram (EMG) signal classification. Similar to [24], they also extracted approximate entropy from direct signals with no transformation. Also, other entropy measures such as sample entropy [27–29], Shannon entropy [30, 31] and Rényi entropy [32, 33] are used in many studies.

Fractal dimension based feature are also got wide attention of researchers in recent years. The fractal dimensions are better measures of complexity of a nonlinear or non-stationary data [35]. In this work we extracted three different fractal dimension features such as Petrosian fractal dimension, Higuchi fractal dimension and Katz fractal dimension. These measures are used in various EEG related studies in the literature. In a study of drowsiness detection [36], authors extracted Petrosian and Higuchi fractal dimensions from EEG time domain signals. Similarly in another work, Acharya et al. [37] extracted Katz fractal dimension with other features for the classification of various sleep stages. We have also extracted one of the newly introduced feature, namely exponential energy by Fasil and Rajesh [26]. Some of the statistical features (mean, standard deviation and energy) are also tested in this work.

5. Experiments and results

In this work, seizure EEG signals and non-seizure EEG signals are classified by decomposing the EEG signal into IMFs using empirical mode decomposition. The frequencies beyond 60 Hz are irrelevant in the EEG analysis due to the non-availability of proper information in higher frequencies [34]. A sixth order butter-worth filter is used to remove frequencies beyond 60 Hz. The signals are further segmented into 10 non-overlapping segments. Empirical mode decomposition is applied on the segmented EEG signals and first four IMFs are obtained. Feature are extracted from four IMFs and averaged across the segments. Support vector

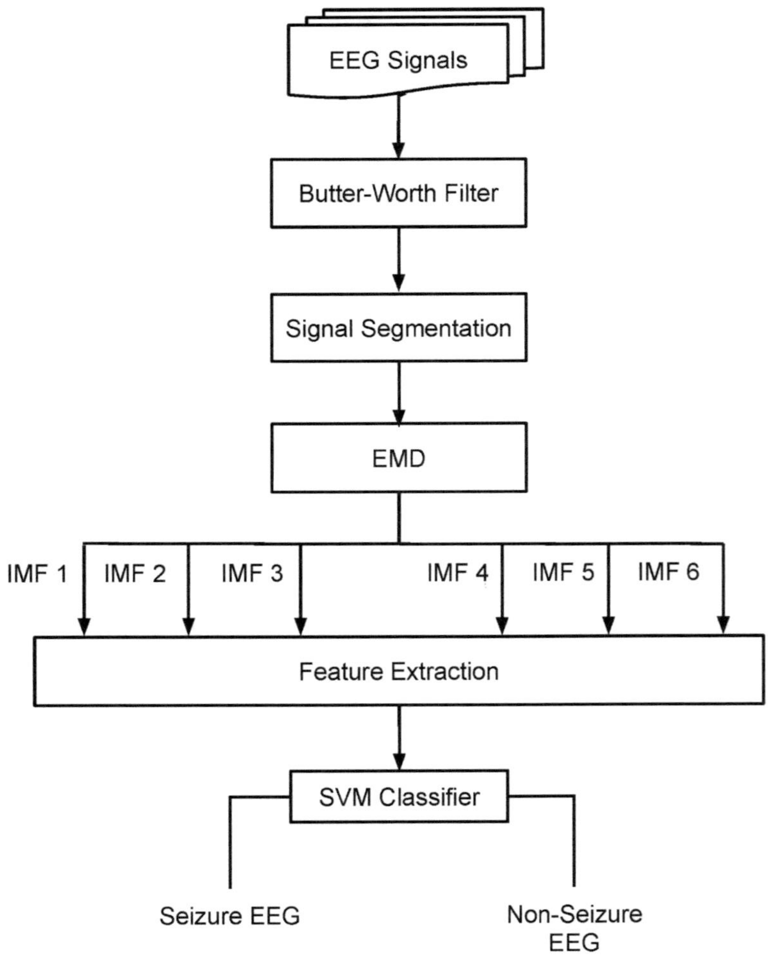

Figure 1.
Block diagram of the proposed seizure classification method.

machine with RBF kernel is used for the classification task. An overall diagram of the work is given in **Figure 1**.

The empirical mode decomposition produces six IMFs in total, though we have considered only first four IMFs. The reason behind this selection procedure is the non-availability of useful information in last IMFs. In this work we have extracted various features such as approximate entropy (ApEn), sample entropy (SmEn), Shannon entropy (ShEn), Rényi entropy (RnEn), Petrosian fractal dimension, Higuchi fractal dimension, Katz fractal dimension, exponential energy from four IMFs and statistical feature (mean, standard deviation and energy).

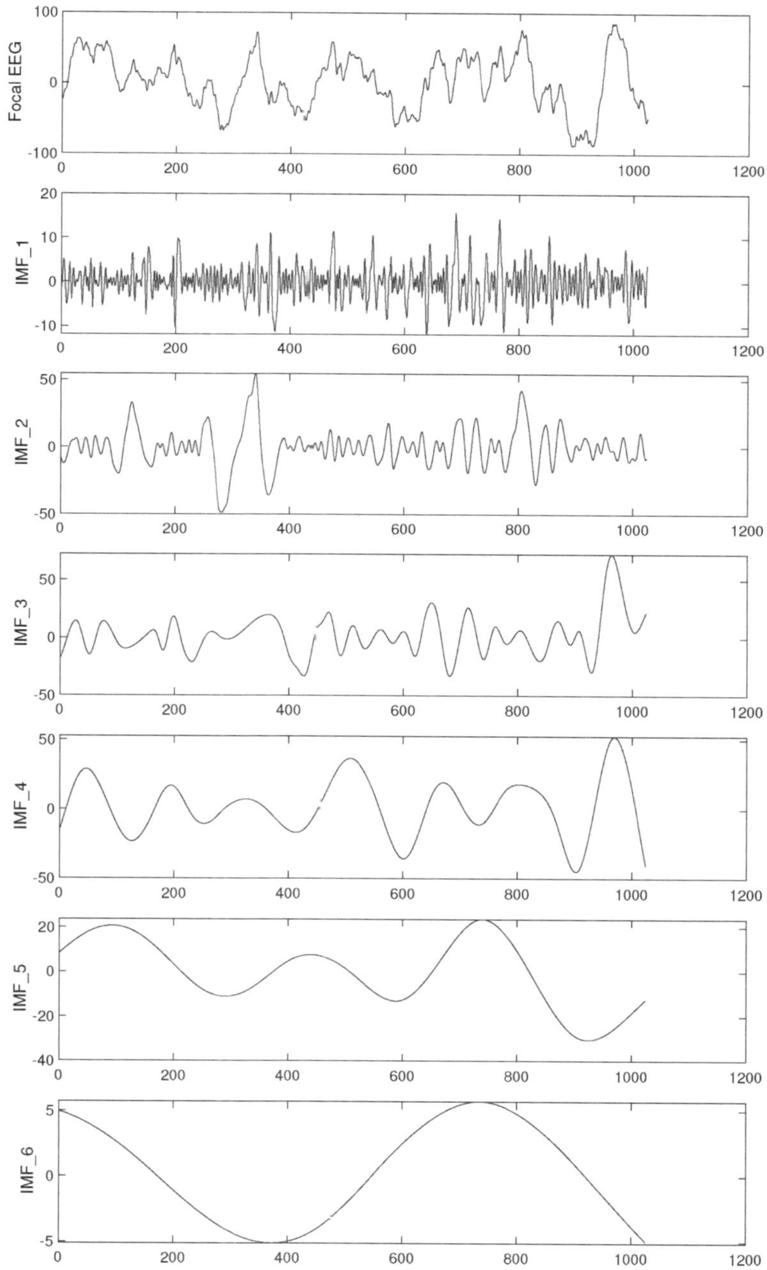

Figure 2.
Focal EEG signals and six IMFs obtained from focal EEG signal.

Sample focal EEG signal and IMFs obtained from focal EEG signals are shown in **Figure 2**. Similarly sample non-focal EEG signal and IMFs obtained from non-focal EEG signals are shown in **Figure 3**.

Each record in the dataset contains a pair of signals denoted as '*x*' and '*y*'. EMD applied on both signal separately and total 8 IMFs are obtained (4 from '*x*' and 4 from '*y*'). To investigate the ability of each feature to classify seizures, experiments are conducted on all features individually. In classification task, the capacity of RBF kernel in support vector machine is already proved in various seizure studies [26, 34]. In this work we have used RBF kernel in support vector machine for the classification task.

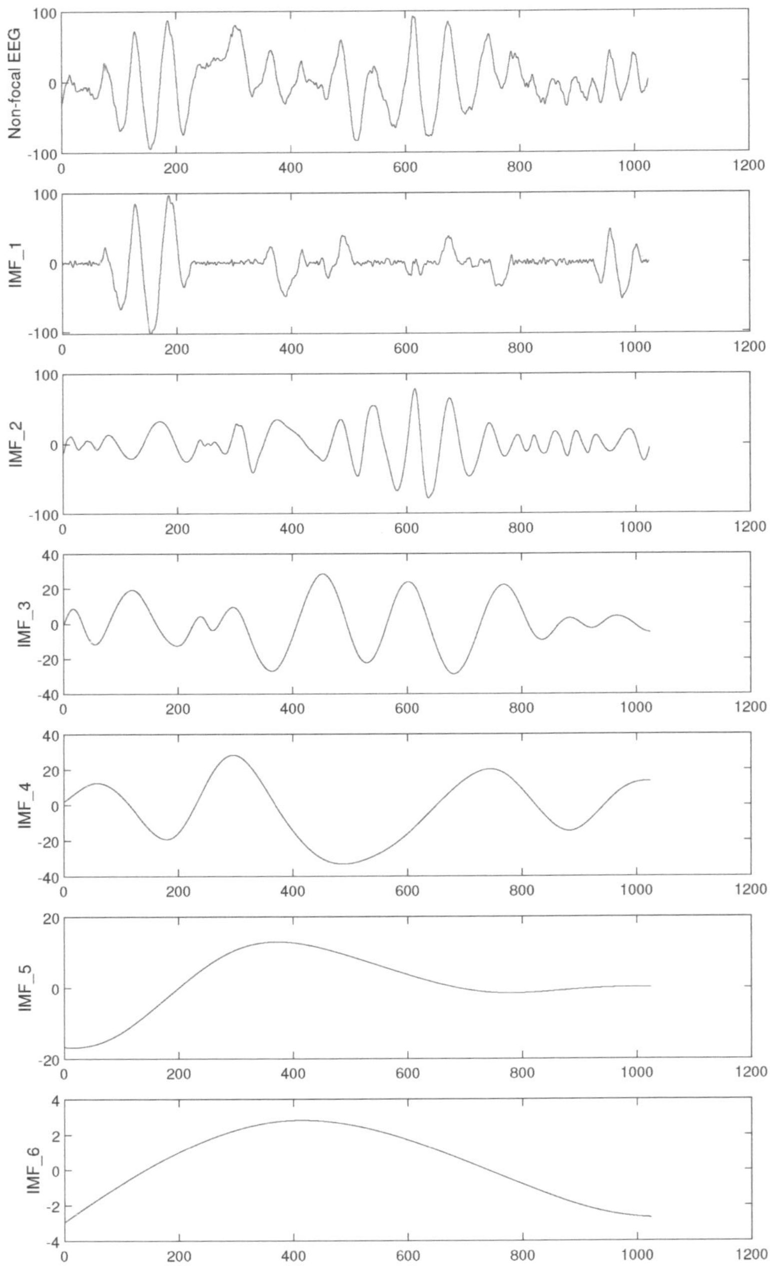

Figure 3.
Non-focal EEG signals and six IMFs obtained from non-focal EEG signal.

Feature	Accuracy (%)	Sensitivity (%)	Specificity (%)
Shannon entropy	58	26	90
Statistical features	68	54	82
Higuchi fractal dimension	71	72	70
Sample entropy	73	83	60
Approximate entropy	75	82	68
Petrosian fractal dimension	75	68	82
Rényi entropy	79	82	76
Katz fractal dimension	80	88	72
Exponential energy	84	84	84

Table 1.
Results of various features extracted from the IMFs.

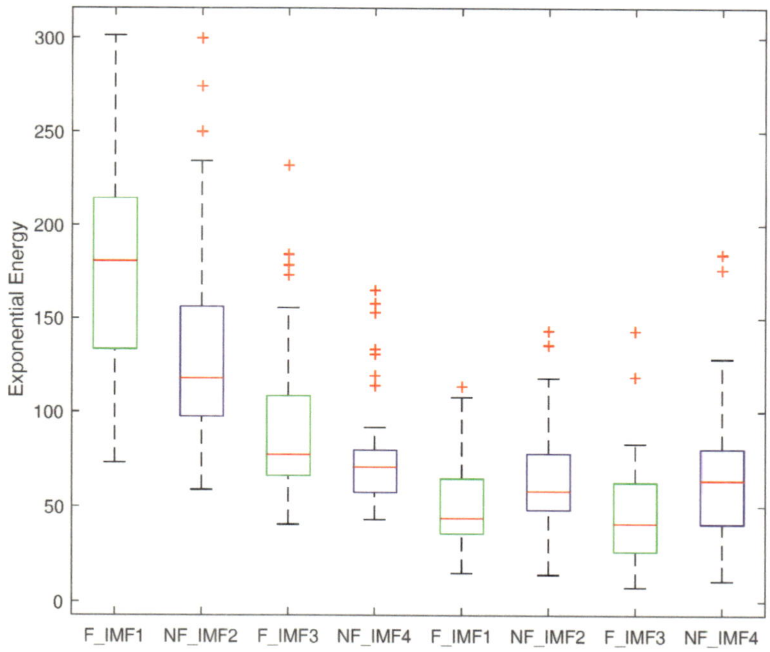

Figure 4.
Boxplot of the extracted exponential energy feature of four IMFs of focal (green color box, labeled as F_IMF) and non-focal (blue color box, labeled as NF_IMF) EEG signals.

We have used k-fold cross-validation with $k = 5$ for testing the ability of extracted features from IMFs. The results of the experiments are calculated with three benchmark measures, such as accuracy, sensitivity and specificity. The measured results are tabulated in **Table 1**.

The results in **Table 1** indicates that the features extracted from the IMFs of empirical mode decomposition gives promising results. Among the tested features, exponential energy feature provided better accuracy with 84%. A box-plot of extracted exponential energy is shown in **Figure 4**. Katz Fractal Dimension also provides better accuracy of 80% followed by Rényi entropy with 79%. Statistical features and Shannon entropy gives less accuracy out of all. It is noted that Shannon entropy giving very low sensitivity value and very high specificity, which indicates that more number of seizure signals are miss-classified as non-seizure signals.

6. Discussion

The study of EEG signals using empirical mode decomposition (EMD) gives an insight into the effectiveness of EMD method to analyze EEG signal for seizure classification. The features (includes four types of entropy features, three types of fractal dimensions, statistical features and exponential energy) considered in this work, produces better classification accuracy when it is extracted from decomposed IMFs.

Empirical mode decomposition method decomposes the signals into various intrinsic mode functions (IMFs). Since, IMFs carries more detailed information of a signal, the features extracted from these IMFs leads to better classification.

Similar to EMD, discrete wavelet transformation (DWT) is a method, which decomposes the signal into various sub-bands [38–40, 42]. Many EEG related studies used DWT method for various analysis. Li et al. [41] combined DWT method with envelope analysis for the effective feature extraction to classify epileptic signal. In another work, Kumar et al. [42] extracted fuzzy entropy from the sub-bands of DWT for seizure detection. Similarly Liu et al. [43], Mohammadi et al. [44] and Silveira et al. [45] also used DWT method to analyze EEG signals for various purposes. Though, EMD is more better than the DWT method.

A comparison of EMD method with DWT is also carried out in this work. The same features which are extracted from the IMFs are also extracted from the DWT sub-bands and classified with same classifier. A bar chart of the comparison of classification accuracy is given in **Figure 5**. The comparison results show that the EMD based feature produces better classification results than DWT based features. EMD based method produced an average accuracy of 73.66%. In case of DWT the average accuracy is 68%. Although, DWT methods shows a slight improvements in results for approximate entropy and Shannon entropy features.

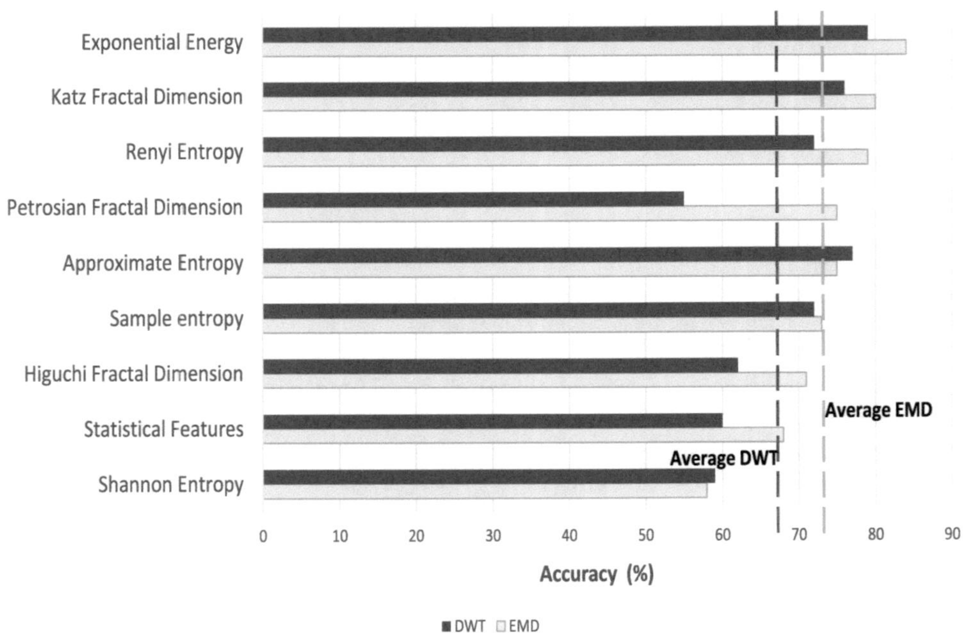

Figure 5.
A comparison of classification accuracy between empirical mode decomposition (EMD) and discrete wavelet transform (DWT). Red dashed vertical line indicates the average accuracy of all DWT features and green dashed vertical line indicates the average accuracy of all EMD features.

Among various entropy features, EMD-Rényi combination (79% accuracy) provides higher classification accuracy. Approximate entropy extracted from IMFs produced an accuracy of 75%. Shannon entropy with EMD is not a good choice of feature for epileptic seizure detection. The classification accuracy produced by Shannon entropy is only 58%. Complexity of EEG data is the reason for less percentage of accuracy.

Three fractal dimensions (Petrosian fractal dimension, Higuchi fractal dimension, Katz fractal dimension) used in this work also produce promising results when they are extracted from IMFs. In this study, EMD based Katz fractal dimension produces higher (80%) classification accuracy than Petrosian (75%) fractal dimension and Higuchi (71%) fractal dimension.

EMD based statistical features did not produce promising results for classification of epileptic EEG signals. But the results are comparatively better than the features from time domain and DWT domain. The highest classification accuracy (84%) reported in this study is with newly introduced exponential energy feature by Fasil and Rajesh [26]. Exponential energy feature utilizes the detailed information available in IMFs to classify epileptic EEG signals effectively. The achieved results show the effectiveness of empirical mode decomposition (EMD) as major step in epilepsy classification.

7. Conclusions

The scope of the empirical mode decomposition of EEG signals in effectual classification of seizure is studied in this work. Four intrinsic mode functions (IMFs) are obtained by applying EMD on filtered EEG signals. Widely used features such as entropy features, fractal dimension features, statistical features and exponential energy features are extracted and its discriminating power is studied. SVM with RBF kernel is used for the classification task. Exponential energy feature provided better results for the seizure classification.

Seizure identification is a challenging and risk bearing activity, which require better accuracy. In future, authors will concentrate on improving the results by incorporating other signal transformation methods with EMD.

Conflict of interest

Authors declare no conflict of interest.

Author details

Fasil OK* and Reghunadhan Rajesh
Central University of Kerala, Kasaragod, Kerala, India

*Address all correspondence to: fasilok92@gmail.com

References

[1] Falco-Walter JJ, Scheffer IE, Fisher RS. The new definition and classification of seizures and epilepsy. Epilepsy Research. 2018;**139**:73-79

[2] Saab ME, Gotman J. A system to detect the onset of epileptic seizures in scalp EEG. Clinical Neurophysiology. 2005;**116**(2):427-442

[3] Tzimourta KD, Tzallas AT, Giannakeas N, Astrakas LG, Tsalikakis DG, Tsipouras MG. Epileptic seizures classification based on long-term EEG signal wavelet analysis. In: Precision Medicine Powered by Health and Connected Health. Singapore: Springer; 2018. pp. 165-169

[4] Annegers JF, Coan SP. The risks of epilepsy after traumatic brain injury. Seizure. 2000;**9**(7):453-457

[5] Gupta A, Singh P, Karlekar M. A novel signal modeling approach for classification of seizure and seizure-free EEG signals. IEEE Transactions on Neural Systems and Rehabilitation Engineering. 2018;**26**(5):925-935

[6] Kumar TS, Kanhangad V, Pachori RB. Classification of seizure and seizure-free EEG signals using local binary patterns. Biomedical Signal Processing and Control. 2015;**15**:33-40

[7] Huang NE, Shen Z, Long SR, Wu MC, Shih HH, Zheng Q, et al. The empirical mode decomposition and the Hilbert spectrum for nonlinear and non-stationary time series analysis. Proceedings of the Royal Society of London, Series A: Mathematical, Physical and Engineering Sciences. 1998;**454**(1971):903-995

[8] Nunes JC, Bouaoune Y, Delechelle E, Niang O, Bunel P. Image analysis by bidimensional empirical mode decomposition. Image and Vision Computing. 2003;**21**(12):1019-1026

[9] Zeng W, Yuan C, Wang Q, Liu F, Wang Y. Classification of gait patterns between patients with Parkinsons disease and healthy controls using phase space reconstruction (PSR), empirical mode decomposition (EMD) and neural networks. Neural Networks. 2019;**111**: 64-76

[10] Hasan NI, Bhattacharjee A. Deep learning approach to cardiovascular disease classification employing modified ECG signal from empirical mode decomposition. Biomedical Signal Processing and Control. 2019;**52**: 128-140

[11] Mi X, Liu H, Li Y. Wind speed prediction model using singular spectrum analysis, empirical mode decomposition and convolutional support vector machine. Energy Conversion and Management. 2019;**180**: 196-205

[12] Thilagaraj M, Rajasekaran MP. An empirical mode decomposition (EMD)-based scheme for alcoholism identification. Pattern Recognition Letters. 2019;**125**:133-139

[13] Hadoush H, Alafeef M, Abdulhay E. Automated identification for autism severity level: EEG analysis using empirical mode decomposition and second order difference plot. Behavioural Brain Research. 2019;**362**: 240-248

[14] Ghritlahare R, Sahu M, Kumar R. Classification of two-class motor imagery EEG signals using empirical mode decomposition and Hilbert-Huang transformation. In: Computing and Network Sustainability. Singapore: Springer; 2019. pp. 375-386

[15] Gaur P, Pachori RB, Wang H, Prasad G. A multi-class EEG-based BCI classification using multivariate

empirical mode decomposition based filtering and Riemannian geometry. Expert Systems with Applications. 2018; **95**:201-211

[16] Martis RJ, Acharya UR, Tan JH, Petznick A, Yanti R, Chua CK, et al. Application of empirical mode decomposition (EMD) for automated detection of epilepsy using EEG signals. International Journal of Neural Systems. 2012;**22**(06):1250027

[17] Zhu G, Li Y, Wen PP, Wang S, Xi M. Epileptogenic focus detection in intracranial EEG based on delay permutation entropy. In: AIP Conference Proceedings; AIP; Vol. 1559, No. 1; 2013. pp. 31–36

[18] Sharma R, Pachori RB, Gautam S. Empirical mode decomposition based classification of focal and non-focal EEG signals. In: 2014 International Conference on Medical Biometrics; IEEE; 2014. pp. 135–140

[19] Sharma R, Pachori R, Acharya U. An integrated index for the identification of focal electroencephalogram signals using discrete wavelet transform and entropy measures. Entropy. 2015;**17**(8): 5218-5240

[20] Chen W, Wang Z, Xie H, Yu W. Characterization of surface EMG signal based on fuzzy entropy. IEEE Transactions on Neural Systems and Rehabilitation Engineering. 2007;**15**(2): 266-272

[21] Lv Z, Wu XP, Li M, Zhang D. A novel eye movement detection algorithm for EOG driven human computer interface. Pattern Recognition Letters. 2010;**31**(9):1041-1047

[22] Li T, Zhou M. ECG classification using wavelet packet entropy and random forests. Entropy. 2016;**18**(8):285

[23] Pincus SM. Approximate entropy as a measure of system complexity. Proceedings of the National Academy of Sciences. 1991;**88**(6):2297-2301

[24] Holzinger A, Stocker C, Bruschi M, Auinger A, Silva H, Gamboa H, Fred A. On applying approximate entropy to ECG signals for knowledge discovery on the example of big sensor data. In: International Conference on Active Media Technology; Springer: Berlin, Heidelberg; 2012. pp. 646–657

[25] Ahmad SA, Chappell PH. Surface EMG classification using moving approximate entropy. In: 2007 International Conference on Intelligent and Advanced Systems; IEEE. 2007. pp. 1163–1167

[26] Fasil OK, Rajesh R. Time-domain exponential energy for epileptic EEG signal classification. Neuroscience Letters. 2019;**694**:1-8

[27] Richman JS, Moorman JR. Physiological time-series analysis using approximate entropy and sample entropy. American Journal of Physiology. Heart and Circulatory Physiology. 2000;**278**(6):H2039-H2049

[28] Jie X, Cao R, Li L. Emotion recognition based on the sample entropy of EEG. Bio-medical Materials and Engineering. 2014;**24**(1):1185-1192

[29] Absolo D, Hornero R, Espino P, Alvarez D, Poza J. Entropy analysis of the EEG background activity in Alzheimer's disease patients. Physiological Measurement. 2006;**27**(3):241

[30] Cao C, Slobounov S. Application of a novel measure of EEG non-stationarity as Shannon-entropy of the peak frequency shifting for detecting residual abnormalities in concussed individuals. Clinical Neurophysiology. 2011;**122**(7): 1314-1321

[31] Yulmetyev RM, Emelyanova NA, Gafarov FM. Dynamical Shannon entropy and information Tsallis entropy

in complex systems. Physica A: Statistical Mechanics and its Applications. 2004;**341**:649-676

[32] Liang Z, Wang Y, Sun X, Li D, Voss LJ, Sleigh JW, et al. EEG entropy measures in anesthesia. Frontiers in Computational Neuroscience. 2015;**9**:16

[33] Inuso G, La Foresta F, Mammone N, Morabito FC. Brain activity investigation by EEG processing: Wavelet analysis, kurtosis and Renyi's entropy for artifact detection. In: 2007 International Conference on Information Acquisition; IEEE; 2007. pp. 195–200

[34] Fasil OK, Rajesh R, Thasleema TM. Influence of differential features in focal and non-focal EEG signal classification. In: 2017 IEEE Region 10 Humanitarian Technology Conference (R10-HTC); IEEE; 2017. pp. 646–649

[35] Accardo A, Affinito M, Carrozzi M, Bouquet F. Use of the fractal dimension for the analysis of electroencephalographic time series. Biological Cybernetics. 1997;**77**(5): 339-350

[36] Mardi Z, Ashtiani SN, Mikaili M. EEG-based drowsiness detection for safe driving using chaotic features and statistical tests. Journal of Medical Signals and Sensors. 2011;**1**(2):130

[37] Acharya R, Faust O, Kannathal N, Chua T, Laxminarayan S. Non-linear analysis of EEG signals at various sleep stages. Computer Methods and Programs in Biomedicine. 2005;**80**(1): 37-45

[38] Ocak H. Automatic detection of epileptic seizures in EEG using discrete wavelet transform and approximate entropy. Expert Systems with Applications. 2009;**36**(2):2027-2036

[39] Kumar Y, Dewal ML, Anand RS. Epileptic seizures detection in EEG using DWT-based ApEn and artificial neural network. Signal, Image and Video Processing. 2014;**8**(7):1323-1334

[40] Guo L, Rivero D, Seoane JA, Pazos A. Classification of EEG signals using relative wavelet energy and artificial neural networks. In: Proceedings of the First ACM/SIGEVO Summit on Genetic and Evolutionary Computation; ACM; 2009. pp. 177–184

[41] Li M, Chen W, Zhang T. Classification of epilepsy EEG signals using DWT-based envelope analysis and neural network ensemble. Biomedical Signal Processing and Control. 2017;**31**: 357-365

[42] Kumar Y, Dewal ML, Anand RS. Epileptic seizure detection using DWT based fuzzy approximate entropy and support vector machine. Neurocomputing. 2014;**133**:271-279

[43] Liu Y, Zhou W, Yuan Q, Chen S. Automatic seizure detection using wavelet transform and SVM in long-term intracranial EEG. IEEE Transactions on Neural Systems and Rehabilitation Engineering. 2012;**20**(6): 749-755

[44] Mohammadi Z, Frounchi J, Amiri M. Wavelet-based emotion recognition system using EEG signal. Neural Computing and Applications. 2017;**28**(8):1985-1990

[45] Silveira TD, Kozakevicius AD, Rodrigues CR. Drowsiness detection for single channel EEG by DWT best m-term approximation. Research on Biomedical Engineering. 2015;**31**(2): 107-115

Chapter 6

Detection of Epileptic Seizure Using STFT and Statistical Analysis

Furkan Kalin, T. Cetin Akinci, Deniz Türkpence, Serhat Seker and Ufuk Korkmaz

Abstract

In this study, EEG data from two volunteer individuals, a healthy individual and a patient with epilepsy, were investigated with two different methods in order to distinguish healthy and patient individuals from each other. The data were obtained from a healthy individual and from a patient with epilepsy at the time of epileptic seizure and of seizure-free interval. The data are those of which validity and reliability were proven and were supplied from the data bank records of University Hospital of Bonn in Germany. In the study, the statistical parameters of the collected data were calculated, then the same data were analysed using short-time Fourier transform (STFT) method, and then they were compared. Both statistical parameter results and spectrum analysis results are compatible with each other, and they can successfully detect healthy individuals and epileptic patients at the time of epileptic seizure and seizure-free interval. In this sense, the results were mathematically highly compatible, which offers significant information for the diagnosis of the disease. In the analysis, the variance values were determined as 253.203 for the healthy individual, 806.939 for the patient at seizure-free interval and 6985.755 for that patient at the time of seizure. Accordingly, standard deviation can be said to be quite distinctive in the designation of values. The frequencies of all three cases resulted in 0, 0–5 and 0–20 Hz, respectively, as a result of conducted STFT analysis, which is quite consistent with the results of the statistical analysis parameters.

Keywords: electroencephalogram, statistical analysis, epilepsy, STFT, seizure

1. Introduction

Temporary clinical conditions, including loss of consciousness, sensory, autonomic and mental disorders, arising from excessive electrical discharges in the nerve cells in the brain, with certain intervals are called as seizure. The condition which becomes chronic with the repetition of these seizures is called as epilepsy. Epilepsy is a chronic disorder that affects the brain and that can be encountered in people of all age groups. It is a neurological disease, most commonly seen in childhood and adolescence periods, and is the second most common disease in adults, followed by brain vessel diseases [1–3]. According to the World Health Organization data, 50 million people around the world are patients with epilepsy [4, 5]. EEG is also used as an auxiliary diagnostic method in the diagnosis of epilepsy, in addition to clinical information, and EEG analysis is performed on patients who are considered to have epileptic seizures [6–8]. Mathematical and spectral methods are used very effectively for the diagnosis of the disease during the analysis of EGG data [9–11].

EEG method establishes the basis of epilepsy science, and its history dates back to the 1940s and is used since then. In principle, it is based on the recording of fluctuations of electrical activity of neurons in the brain, and the main contributions of EEG for epileptic cases can be summarized as follows: supports a clinically identified diagnosis; is used as a confirmatory test; helps to make diagnosis correctly; directly and indirectly identifies seizure type and epilepsy syndrome, together with some findings; and informs about the location of focus [12].

Delta (δ) waves are those with frequencies of 1–4 Hz and amplitudes of 20–400 μV. They are seen in cases when the brain has very low activity, such as deep sleep, general anesthesia, immune system, natural recovery.

Theta (θ) waves are those with frequencies of 5–7 Hz and amplitudes of 5–100 μV. They are seen in cases when the brain has low activity, such as sleep with dream, middle anesthesia, stress, emotional commitment.

Alpha (α) waves are those with frequencies of 8–13 Hz and amplitudes of 2–10 μV. They are seen in cases when awake individuals are physically and mentally full resting, there is no any external stimulant, in relaxed positions and when eyes are closed. They are most prominently observed in records obtained from the occipital region.

Beta (β) waves are those with frequencies of 14–30 Hz and amplitudes of 1–5 μV. They are seen in cases, including focused attention, mental work, problem solving, memory, sensory information processing, rapid eye movements phase of sleep [13–16].

The EEG signals used in this study are registered at the University Hospital of Bonn in Germany [17]. The dataset consists of five subsets (denominated as A, B, C, D and E) that are recorded with the same 128-channel amplifier system and 12-bit analog-to-digital converter. Each of the subsets contains 100 segments with a sampling frequency of 173.61 Hz and a duration of 23.6 s, i.e. 4096 sample points; the corresponding frequency bandwidth is 86.8 Hz. Subsets B, D and E were analysed in this study. While EEG samples in set B were obtained from five healthy volunteers via external surface electrodes, for closed eye condition, set D consisted of EEG segments recorded from patients with epilepsy using intracranial electrodes to monitor epileptic activity, obtained at the time without seizure. Set D data were obtained from epileptic area, and they were recorded. Set E contains EEG data, obtained from patients with epilepsy, recorded at the time of seizure. Strip electrodes were used while recording set E data.

2. Statistical and mathematical background

EEG signals are not deterministic. Since EEG signals do not have a specific shape as electrocardiogram (ECG) signals do, statistical and parametric methods are used in the analysis of EEG signals [18–20]. Spectral methods are used for the classification and characterization of EEG signals [21–35].

2.1 Statistical analysis methods

Statistical parameters are used to obtain necessary properties, in most of the analyses performed in time domain. Although the mean and median values of signal are expected to result in pretty near zero when the signals have a periodical and sinusoidal structure, these values can get away from zero by taking positive or negative values in non-periodical signals. In the analyses, the most basic mean value, μ, and the standard deviation, σ, can produce distinctive results on non-periodic signals [19]. For a given data set {xi}, these are defined as follows:

$$\mu = \frac{1}{N}\sum_{i=1}^{N} x_i \tag{1}$$

$$\sigma = \sqrt{\frac{1}{N}\sum_{i=1}^{N}(x_i - \mu)^2}. \tag{2}$$

where N is the number of the data points.

Knowing the standard deviation value for a given series of numbers and understanding this concept mean knowing to what extent this series is distributing around an average. The bigger standard deviation indicates that datapoints get further away from the average; a small standard deviation indicates that data points form more close groups around the average.

In practice, data often correspond to normal probability distribution (Gaussian), which is actually due to the central limit theorem. According to the central limit theorem, the sum of random variables, which are independent of each other and all have the same distribution, tends to follow a normal distribution at the limit. Here, skewness (α) and kurtosis (β), two functions obtained from the Gaussian distribution, are given in the following Equations [20]:

$$\alpha = \frac{\left[\frac{1}{N}\sum_{i=1}^{N}(x_i - \mu)^3\right]}{\sigma^3} \tag{3}$$

$$\beta = \frac{\left[\frac{1}{N}\sum_{i=1}^{N}(x_i - \mu)^4\right]}{\sigma^4} \tag{4}$$

Here, when α equals zero, it means a perfect normal distribution, and when α takes negative or positive values, it means symmetry is deflected towards the right or left side. In statistical calculations, if the skewness is negative, the tail of the curve will extend to the left, and the distribution will densify on the right side of the graph. If the skewness is positive, the tail of the curve will extend to the right, and the distribution will densify on the left side of the graph. The kurtosis (β) is very close to 3 for the normal distribution. These statistical parameters can be used to quickly check the changes in the statistical behaviour of a signal [18–20].

2.2 Fourier transform and STFT

The Fourier transform (FT) method is one of the most effective methods used to process signals, in order to obtain information in the signal. In the Fourier transform method, a signal is expressed as the sum of the fundamental cosine and sinus components at different amplitudes, frequencies, and phases. The tabulation of each component with its frequency and amplitude provides convenience during data processing through computers. The equations for Fourier transform are given below in Eq. (5) and Eq. (6) [23, 26, 29, 30, 33]:

$$f(x) = \frac{1}{\sqrt{2\pi}}\int_{-\infty}^{\infty} F(k)\, e^{ikx}\, dk \tag{5}$$

$$F(k) = \frac{1}{\sqrt{2\pi}}\int_{-\infty}^{\infty} F(x)\, e^{-ikx}\, dx \tag{6}$$

Similarly, based on the Fourier transform method, short-time Fourier transform (STFT) and spectrogram were developed by Gabor in 1946. This method most clearly reveals time and frequency localization [17]. STFT provides very decisive results in the analysis of signals. Here, a x(t) signal is used in a fixed window size and in frequency resolution. To define the STFT, let us consider a signal x(t) with assumption that it is stationary when it is windowed through a fixed dimension window g(t), cantered at time location τ. The Fourier transform of the windowed signal yields the STFT [23, 25, 26, 33–35].

$$STFT\{x(t)\} \equiv X(\tau,f) = \int_{-\infty}^{\infty} x(t)g(t-\tau)e^{-j2\pi ft}dt \tag{7}$$

Similarly, for two-dimensional, discretely timed signals, this time-frequency function (t, f) is given in Eq. (8). Here, window g(t) is chosen; the STFT resolution is fixed over the entire time-frequency plane [23–30]:

$$STFT\{x(n)\} \equiv X(m,f) = \sum_{n=-\infty}^{\infty} x(n)g(n-m)e^{-j\omega n} \tag{8}$$

The spectrogram is given in Eq. (9):

$$\{x(t)\} \equiv |X(\tau,f)|^2 \tag{9}$$

3. Analysis and application

In this study, the data from two different volunteers were used in the analysis of EGG data. One of these individuals is healthy and the other one is a patient with epilepsy. The healthy individual's eyes are closed (**Figure 1a**). The data of the patient with epilepsy were collected when he did not suffer a seizure, and these data were taken from the epileptic area (**Figure 1b**).

When the data of two individuals, healthy and patient with epilepsy, given in **Figure 1**, are examined, it is seen that while the graphical amplitude of the healthy individual changes in the range of 0–40 μV as shown in **Figure 1a**, the graphical amplitude of the patient with epilepsy rises up to 125 μV as shown in **Figure 1b**. As is understood from this graph, the amplitudes of two individuals, healthy and

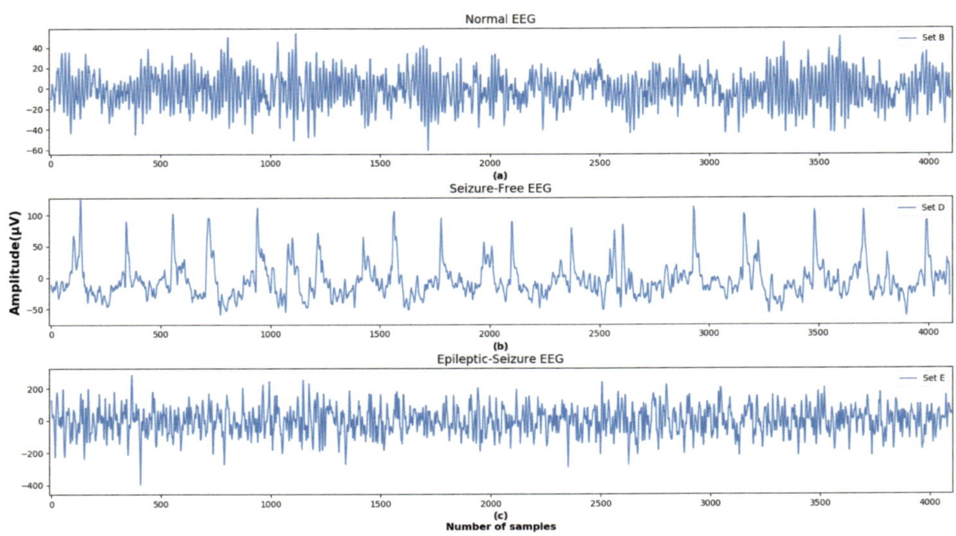

Figure 1.
(a) Healthy (his eyes closed), (b) EEG signal from epileptic area, with seizure-free intervals and (c) EEG signal from the epileptic region at the time of seizure.

	Mean (μ)	Standard deviation (σ)	Variance	Skewness(α)	Kurtosis(β)
Healthy	−1.347	15.912	253.203	−0.027	3.043
Free seizure	−5.24	28.406	806.939	1.437	5.738
Seizure	−2.521	33.580	6985.755	−0.100	3.268

Table 1.
Statistical analysis of healthy individual's and patient with epilepsy's EEG signal.

patient with epilepsy, are sufficient to clearly diagnose the disease. The graphical amplitude gets maximum value despite the absence of seizure as shown in **Figure 1b**. However, the moment when the same individual has a seizure shown in **Figure 1c**), the amplitudes are seen to increase much. These graphs clearly indicate whether the individual is a patient with epilepsy and whether those patients with epilepsy suffer a seizure (**Table 1**).

When the statistical parameters of the data from the healthy individual and of the epileptic area of the patient with epilepsy are examined, it is seen that there are significant differences both in the mean value and in the standard deviation. The mean value yields −1.347 in the healthy individual, −5.24 in the patient with epilepsy and − 2.521 when the patient with epilepsy suffers a seizure. As a result of the analysis, standard deviation values can be said to be quite decisive in the diagnosis of the disease. However, while healthy individual and the patient with epilepsy can be distinguished from each other in histograms, the histogram shows a normal distribution at the time of seizure.

In this context, histogram graphs can only be used in the diagnosis of epilepsy, but not in detecting seizure moments. The histograms of the healthy individual and of the patient with epilepsy are given in **Figures 2** and **3**. **Figure 4** shows histograms of the data of the patient with epilepsy at the time of seizure.

The spectrum analysis of data received from individuals is given in **Figures 5–7**. The frequency values were limited to 50 Hz, considering general characteristics of EEG frequency values. At the spectral analysis of the data of the healthy individual, the frequency values were seen to be about 10 Hz, and this corresponds to the alpha waves. When the spectrogram of the patient with epilepsy was examined at the time of free seizure, it was observed that the frequency values corresponded

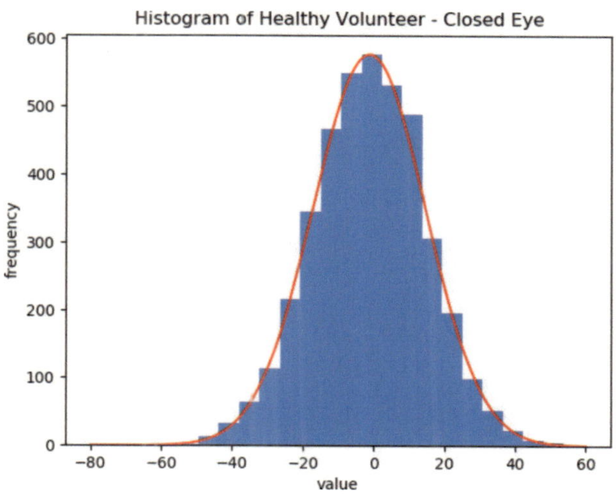

Figure 2.
Set E histogram (healthy volunteer).

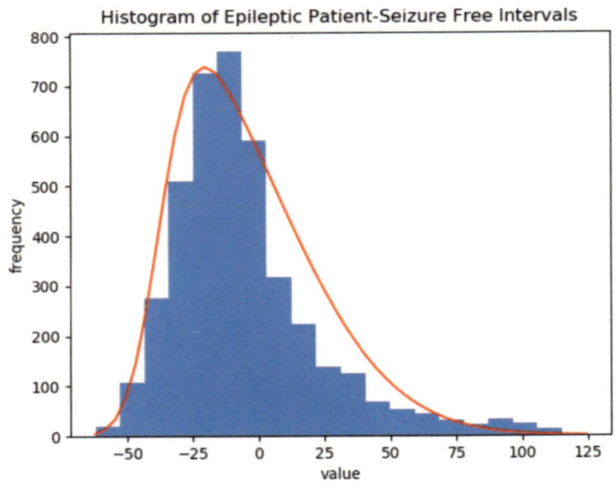

Figure 3.
Set E histogram (seizure-free intervals).

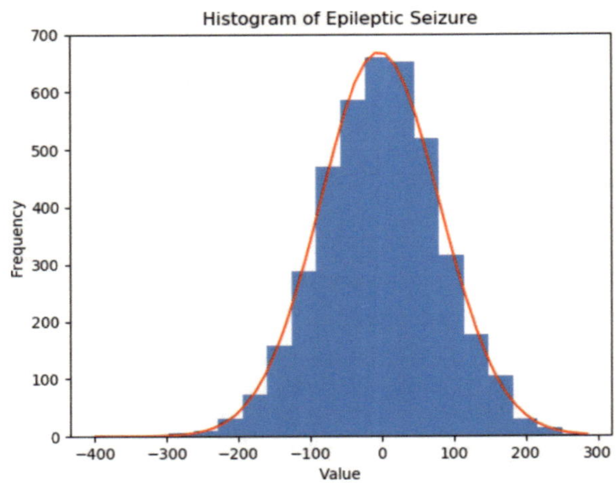

Figure 4.
Set E histogram (epileptic seizure).

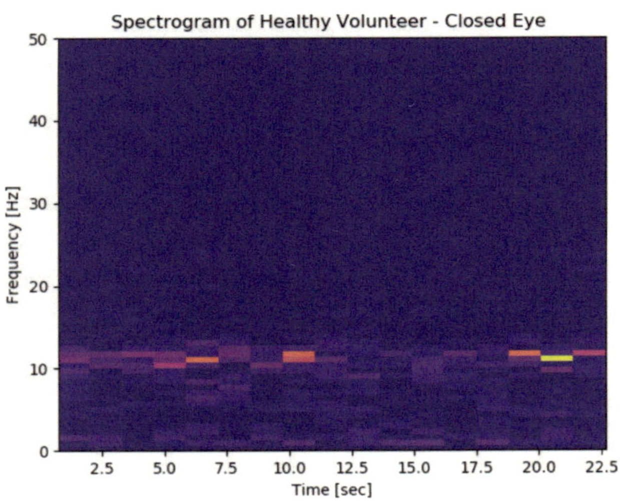

Figure 5.
Spectrogram of healthy individual with closed eye.

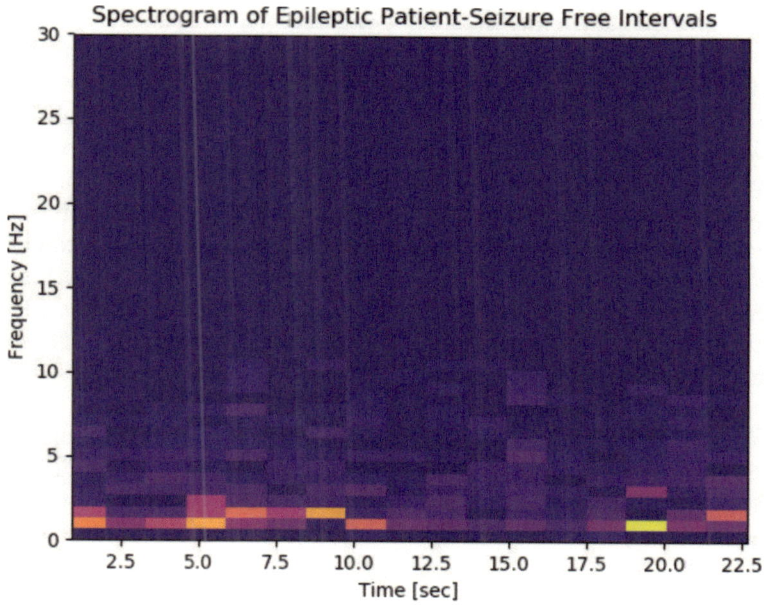

Figure 6.
Spectrogram for seizure-free epileptic patient.

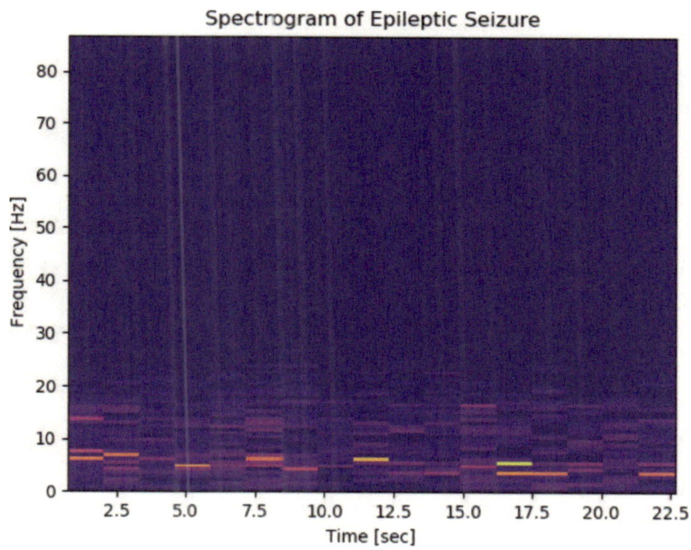

Figure 7.
Spectrogram of a patient with epilepsy during epileptic seizure.

to the range of 0.5–4 Hz (delta wave), and when the data of the same patient were examined at the time of seizure, the frequency values were seen to distribute in the range of 0–20 Hz.

4. Discussion

First of all, it should be noted that the data used in the study provide validity and reliability conditions. In this sense, the data used in this study were used in many articles and obtained from the records of the database of University Hospital of Bonn in Germany [11]. Time-frequency-based techniques have been used in many studies using EEG data [36]. In most of these studies, anomalies in the brain can be determined from the high-frequency difference [36–42]. In this study, the EEG data of individuals with different conditions (non-patient, sick and seizure patient) were analysed. The analyses were compared. This comparison makes it easy to classify the patients. In this study, EEG analysis was approached from different perspectives compared to other studies. Traditionally, basic linear analyses and statistical approaches have been used in time and frequency fields. In this sense, it can be said that the study contains more definite, distinctive results than other analyses. In the literature, the amplitude of the signal, the distance between seizure and non-seizure intervals and the energy ratio of EEG have been investigated. These studies have been used as a criterion for the evaluation of epileptic activity [43–47]. Today, many mathematical methods are used in the analysis of EEG data [48–50]. Data collection systems are constantly changing with the developing technology. In the future, it is predictable that the data will be made by remote sensing. Furthermore, the analysis of EEG data with artificial intelligence methods can be developed as a tool. With this tool, neurofeedback applications can be considered as the most important method in the treatment and development.

5. Conclusions

In this study, the data from healthy individual and from the patient with epilepsy were examined. The collection and analysis of data of the patient with epilepsy at both the time of seizure and of seizure-free interval are important for the diagnosis of the disease. In this study, first of all, statistical analyses were performed, and as a result of the analysis, the mean and standard deviation values of the healthy individual and the patient with epilepsy suggest very decisive results. **Figure 8** shows the comparison of the histograms of the individuals.

Variance, one of the statistical parameters, also produces meaningful results in distinguishing patient and healthy individuals. In this study, variance value yields 253.203 for the healthy individual, 806.939 for the patient with epilepsy at

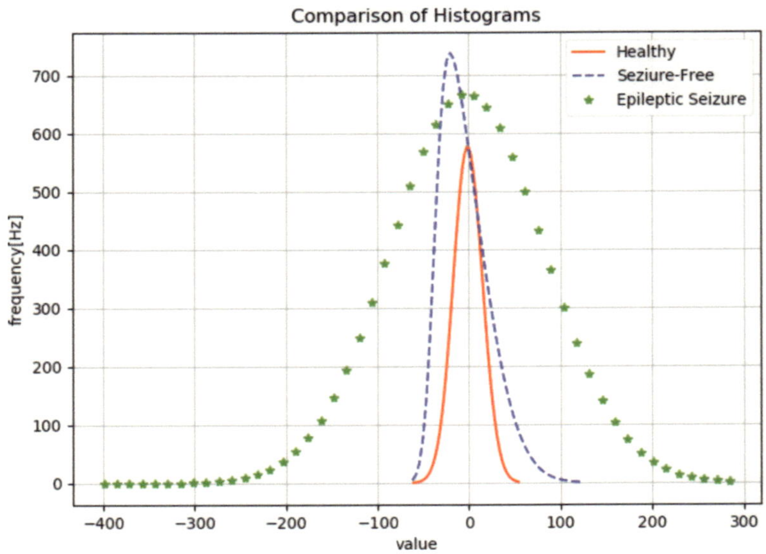

Figure 8.
Comparison of the histograms.

seizure-free interval and 6985.755 for that patient at the time of seizure. Other statistical parameters also revealed very clear results in distinguishing patient and healthy individuals. However, the results of the STFT analyses support the statistical parameters. In this study, the frequency values of the healthy individual were seen to distribute around 10 Hz, in the range of 0–5 Hz for the patient with epilepsy at seizure-free interval and in the range of 0–20 Hz for that patient at the time of seizure. Accordingly, it can be said that the statistical and spectral analyses made are quite decisive for the diagnosis of the disease.

Author details

Furkan Kalin, T. Cetin Akinci*, Deniz Türkpence, Serhat Seker and Ufuk Korkmaz
Department of Electrical Engineering, Istanbul Technical University, Istanbul, Turkey

*Address all correspondence to: akincitc@itu.edu.tr

References

[1] Nathan SK, Brahme IS, Kashkoush AI, Anetakis K, Jankowitz BT, Thirumala PD. Risk factors for in-hospital seizures and new-onset epilepsy in coil embolization of aneurysmal subarachnoid hemorrhage. World Neurosurgery. 2018;**115**:e523-e531

[2] Claassen J, Jette N, Chum F, Green R, Schmidt M, Choi H, et al. Electrographic seizures and periodic discharges after intracerebral hemorrhage. Neurology. 2007;**69**:1356-1365

[3] Lin CL et al. Characterization of perioperative seizures and epilepsy following aneurysmal subarachnoid hemorrhage. Journal of Neurosurgery. 2003;**99**(6):978-985

[4] World Health Organization Epilepsy. Available from: http://www.who.int/news-room/fact-sheets/detail/epilepsy [Accessed: April 2018].

[5] Goldenberg MM. Overview of drugs used for epilepsy and seizures etiology, diagnosis, and treatment. Pharmacy and Therapeutics. 2010;**35**(7):392-415. https://www.ncbi.nlm.nih.gov/pmc/articles/PMC2912003/

[6] Tekeli H, Yasar H, Kendirli MT, Senol MG, Ozdag F, Saracoglu M. The prevalence of epilepsy in young Turkish males. Epilepsi: Journal of The Turkish Epilepsy Society. 2012;**18**(1):1-6

[7] Blume WT. Current trends in electroencephalography. Current Opinion in Neurology. 2001;**14**:193-197

[8] Haas LF. Hans Berger (1873-1941), Richard Caton (1842-1926), and electroencephalography. Journal of Neurol Neurosurg Psychiatry. 2003;**74**(1):9. DOI: 10.1136/jnnp.74.1.9

[9] Singh P, Joshi SD, Patney RK, Saha K. Fourier-based feature extraction for classification of EEG signals using EEG rhythms. Circuits, Systems, and Signal Processing. 2016;**35**(10):3700-3715

[10] Uthayakumar R, Easwaramoorthy D. Epileptic seizure detection in EEG signals using multi fractal analysis and wavelet transform. Fractals. 2013;**21**(2):24. Article ID: 1350011. https://www.worldscientific.com/doi/abs/10.1142/S0218348X13500114

[11] Available from: https://www.massgeneral.org/childhood-epilepsy/overview/brain.aspx [Accessed: May 2018]

[12] Neurology: Clinical textbooks of Istanbul Medical Faculty. 2004. İstanbul: Istanbul University Istanbul Medical Faculty

[13] Britton JW, Frey LC, Hopp JL, et al. Electroencephalography (EEG): An Introductory Text and Atlas of Normal and Abnormal Findings in Adults, Children, and Infants. In: Louis EKS, Frey LC, editors. Chicago: American Epilepsy Society; 2016. pp. 8-76. https://www.ncbi.nlm.nih.gov/books/NBK390345/

[14] Malow BA, Passaro E, Milling C, Minecan DN, Levy K. Sleep deprivation does not affect seizure frequency during inpatient video-EEG monitoring. Neurology. 2002;**59**:1371-1374

[15] Claassen J, Mayer SA, Kowalski RG, Emerson RG, Hirsch LJ. Detection of electrographic seizures with continuous EEG monitoring in critically ill patients. Neurology. 2004;**62**:1743-1748

[16] Al-Kadi MI, Reaz MBI, Ali MAM. Evolution of electroencephalogram signal analysis techniques during anesthesia. Sensors (Basel). 2013;**13**(5):6605-6635

[17] Epileptologie Bonn/Forschung/AG Lehnertz/EEG Data Download. 2018;64:061907. DOI: 10.1103/PhysRevE.64.061907. Retrieved from: http://epileptologie-bonn.de/cms/front_content.php?idcat=193&lang=3&changelang=3

[18] Singh P, Pachori RB. Classification of focal and nonfocal EEG signals using features derived from Fourier-based rhythms. Journal of Mechanics in Medicine and Biology. 2017;**17**(07):1740003. https://www.worldscientific.com/doi/abs/10.1142/S0219519417400036

[19] Kaur J, Kaur A. A review on analysis of EEG signals. IEEE International Conference on Advances in Computer Engineering and Applications. 2015:957-960. DOI: 10.1109/ICACEA.2015.7164844. https://ieeexplore.ieee.org/document/7164844

[20] Pradhan N, Dutt DN. Data compression by linear prediction for storage and transmission of EEG signals. International Journal of Biomedical Computing. 1994;**35**(3):207-217. https://ieeexplore.ieee.org/abstract/document/7164844

[21] Ramadan R, Vasilakos A. Brain computer interface: Control signals review. Neurocomputing. 2017;**223**:26-44

[22] Tzallas AT, Tsipouras MG, Fotiadis DI. Epileptic seizure detection in EEGs using time-frequency analysis. IEEE Transactions on Information Technology in Biomedicine. 2009;**13**(5):703-710

[23] Vaseghi VS. Advanced Signal Processing and Digital Noise Reduction. New York: Wiley; 1996

[24] Stoica P, Moses R. Spectral Analysis of Signals. Upper Saddle River, New Jersey, USA: Prentice Hall; 2005, ISBN 0-13-113956-8. pp. 15-33

[25] Taylor JK. Statistical Techniques for Data Analysis. London: Lewis Publishers; 1990

[26] Gröchenig K. Foundations of Time-Frequency Analysis. Birkhäuser Boston: Springer Science; 2001. pp. 37-58

[27] Shaker MM. EEG waves classifier using wavelet transform and Fourier transform. World Academy of Science Engineering and Technology. 2007;**3**:723-728

[28] Şengür A, Guo Y, Akbulut Y. Time-frequency texture descriptors of EEG signals for efficient detection of epileptic seizure. Brain Informatics. 2016;**3**:101-108

[29] Stoica P, Moses RL. Introduction to Spectral Analysis. Prentice hall: Upper Saddle River, NJ, USA; 1997

[30] Ramos-Aguilar R, Olvera-López JA, Olmos-Pineda I. Analysis of EEG signal processing techniques based on spectrograms. Research in Computing Science. 2017:145;151-162. http://www.rcs.cic.ipn.mx/2017_145/Analysis%20of%20EEG%20Signal%20Processing%20Techniques%20based%20on%20Spectrograms.pdf

[31] Langton C, Levin V. The Intuitive Guide to Fourier Analysis and Spectral Estimation. Ist edition, USA: Mountcastle Company; 2017

[32] Walker JS. Fast Fourier Transform. USA: CRC Press; 1996, ISBN: 9780849371639

[33] Herman RL. An Introduction to Fourier Analysis. NY, USA: Chapman and Hall, CRC Press; 2016

[34] Davis SP, Abrams MC, Brault JW. Fourier Transform Spectrometry 2001. pp. 41-66

[35] Zabidi A et al. Short-time Fourier transform analysis of EEG signal generated during imagined writing.

International Conference on System Engineering and Technology (ICSET). Elsevier, Academic Press; 2012. DOI: 10.1016/B978-0-12-042510-5. X5000-8. ISBN: 978-0-12-042510-5. https://www.sciencedirect.com/book/9780120425105/fourier-transform-spectrometry

[36] Faust O, Acharya UR, Adeli H, Adeli A. Wavelet-based EEG processing for computer-aided seizure detection and epilepsy diagnosis. Bandung, Indonesia: Seizure. 2015;**26**:56-64. https://ieeexplore.ieee.org/document/6339284

[37] Martis RJ, Acharya UR, Lim CM, Mandana KM, Ray AK, Chakraborty C. Application of higher order cumulant features for cardiac health diagnosis using ECG signals. International Journal of Neural Systems. 2013;**23**(4):1350014

[38] Adeli H, Zhou Z, Dadmehr N. Analysis of EEG records in an epileptic patient using wavelet transform. Journal of Neuroscience Methods. 2003;**123**(1):69-87

[39] Litt B, Echauz J. Prediction of epileptic seizures. The Lancet Neurology. 2002;**1**(1):22-30

[40] Päivinen N, Lammi S, Pitkänen A, Nissinen J, Penttonen M, Grönfors T. Epileptic seizure detection: A nonlinear viewpoint. Computer Methods and Programs in Biomedicine. 2005;**79**:151-159

[41] Giannakakis G, Sakkalis V, Pediaditis M, Tsiknakis M. Methods for seizure detection and prediction: An overview. In: Sakkalis V, editor. Modern Electroencephalographic Assessment Techniques. Vol. 91 of Neuromethods. Vol. 2015. New York: Springer. pp. 131-157

[42] Uhlhaas PH, Singer W. High-frequency oscillations and the neurobiology of schizophrenia. Dialogues in Clinical Neuroscience. 2013;**15**(3):301-313

[43] Paul Y. Various epileptic seizure detection techniques using biomedical signals: A review. Brain Informatics. 2018;**5**(2):2-19

[44] Tessy E, Shanir PPM, Manafuddin S. Time domain analysis of epileptic EEG for seizure detection. In: 2016 International Conference on Next Generation Intelligent Systems (ICNGIS). Kottayam, India

[45] Ilaslan MF. The multiresolution wavelet analysis of the brain signals of epilepsy patients. The Journal of Cognitive Systems. 2018;**3**(2):30-33

[46] Meng QF, Chen Y. A novel feature extraction method for epileptic EEG based on degree distribution of complex network. WSEAS Transactions on Information Science and Applications. 2015;**12**:51-60. ISSN: E-ISSN: 2224-3402. Available from: http://www.wseas.org/multimedia/journals/information/2015/a125709-409.pdf

[47] Bajaj V, Pachori RB. Epileptic seizure detection based on the instantaneous area of analytic intrinsic mode functions of EEG signals. Biomedical Engineering Letters. 2013;**3**(1):17-21

[48] Doschoris M, Kariotou F. Electroencephalography-mathematical foundation of electroencephalography, chapter 4, Intech Publication; 2017. pp. 35-54

[49] Rashid M, Bari BS, Sadeque G. Mathematical principals and modeling of EEG signal exploration. In: Sittiprapaporn P, editor. Electroencephalography. IOSR Journal of VLSI and Signal Processing. InTech. 2016;**6**(6):5-9. http://www.who.int/news-room/fact-sheets/detail/epilepsy

[50] Svozilova V, Mezl M. Modeling of the EEG signal. Elektrorevue. 2016;**18**(2):44-51

Chapter 7

Information Processing and Synaptic Transmission

Vito Di Maio and Silvia Santillo

Abstract

The brain is probably the most complex machinery for information processing we can imagine. The amount of data it manages is extremely huge. Any conscious or unconscious event both internal and coming from the environment needs to be perceived, elaborated, and responded with an appropriate action. Moreover, the high-level activities of mind require the connection of logical elaboration, the relationship with past experience (memory), and the transfer of information among different areas of the brain participating to the elaboration of the thought. Almost all brain illnesses or even simple defaults can be related to a corruption of the basic system which manage information in the brain. The main actors in transferring and managing information are the synapses, and then the understanding of the brain information processing cannot disregard the full understanding of the synaptic functionality. In the present chapter, by using as example the most common type of the brain synapse (the glutamatergic synapse), we will present the basic mechanism of synaptic transmission stressing some of the most relevant mechanisms which regulate the transfer and management of information.

Keywords: information processing, synaptic transmission, synaptic integration, neuronal spikes, neuronal modeling, synaptic modeling

1. Introduction

The brain is probably the most complex computational machinery performing in parallel a continuous elaboration of information coming from the environment and from the internal of the body. Undoubtedly, independently from the level of investigation (from the molecular to the neurological and psychological level), almost all the neurosciences deal, directly or indirectly, with the brain information processing and/or its malfunctioning. In this chapter, we will illustrate some basic aspects of information transfer and elaboration showing how much complex is the control of its flow among the neurons.

Neurons share information mainly by the synaptic contacts which they use both to transmit and to receive. The input and output contact among many neurons are the system which operate the neural networks and the whole brain. Synapses are, then, the key points for the information transfer among neurons, but, as we will see in details later, they are also the primarily system of information coding and elaboration. Their activity, in fact, produces the codification of the information by a neuron in form of spike sequences into a sequence of postsynaptic potentials (PSP) which we can define as the first step of the postsynaptic representation of the presynaptic code. If we consider the spike sequence of a presynaptic neuron as the representation of a stimulus, the PSPs produced at the synaptic level will be the synaptic representation of that stimulus. The meaning of stimulus, however, does not only refer to the codification of an environmental stimulation. The spike sequences, in several neurons, are not only the codification of stimuli but participate also to the high-level performances connected to memory recall, thought, reasoning, and so on. Whatever is the role of the spike sequence, it represents an information which, transmitted to other neurons, is translated at the synaptic level in a sequence of PSP. How this will be further recoded into a postsynaptic spike sequence depends on a complex integration of all the inputs arriving to the neuron in a compatible time window.

Although a large effort is spent in the last five decades for its understanding, the way the neurons really code, manipulate, and share information remains a mystery. What seems to be generally accepted is that the code of a neuron, for a given event, is formed by a sequence of elementary bits (spikes) in a given time window. The difficulty in understanding the code for a given stimulus rises because this sequence often seems to be randomly distributed in time (irregular and non-repetitive interspike intervals) also when generated for the same stimulus. So far, two main ideas have been affirmed on the possible nature of the code, and both of them are supported by many strong experimental evidences. According to one of them, the codification of the stimulus occurs in terms of frequency of the spikes in a given time window. Many different time sequences of the spikes can give the same frequency since it depends on the number of spikes given in the chosen time window. The alternative one assumes that the coding is embedded in the precise timing of the spike occurrence.

The difficulty in understanding the relationship between the code generated by neurons in sequences of spikes (either as frequency or precise timing) rises essentially by the lack of the precise knowledge on how the neuron generates spikes thanks to the thousands of synaptic inputs it receives. In turn, this lack of knowledge depends on the still low level of knowledge on how the synapses code the presynaptic information into a sequence of PSP. The understanding of the basic mechanisms of synaptic transmission is fundamental in all fields of neurosciences including the genesis of important brain diseases involving memory impairment and other brain performances as Parkinson [1], Alzheimer [2], and Autism [3]. Not surprisingly then a big effort is spent nowadays worldwide to study synaptic transmission with the most diverse experimental approaches but also with mathematical modeling and computer simulations since, for the structural conformation, not all the properties of the synapses can be unveiled by the experimental approaches.

In the present chapter, we will use the most common type of excitatory synapse in the brain, the glutamatergic synapse, to outline, after a brief simple explanation of its functioning, how many and how complex are the mechanisms controlling the flow of information among the neurons operated by these synapses.

A typical pyramidal neuron of the cortex or of the hippocampus subfields receives thousands of synaptic inputs (3000–30,000) [4–6]. The larger parts of these inputs (80%) are excitatory inputs which use glutamate (Glu) as neurotransmitter. It is then reasonable to assume that these synapses are the most important way of information transfer and elaboration. Probably, the most important regulatory system of the activity of the glutamatergic pyramidal neurons is given by the inhibitory neurons which use the γ-aminobutyric acid (GABA) as neurotransmitter [4–6].

GABAergic synapses represent between 10 and 20% of the synapses inputting on a pyramidal neuron, and they are located in strategic positions on the shaft of the dendritic branches among the excitatory glutamatergic synapses [4–6]. Glutamatergic synapses are normally located on spines (a sort of elongation) protruding from the shaft of the dendritic branches. When activated they produce the so-called excitatory postsynaptic current (EPSC) which is a current which depolarizes the membrane (increases V_m) producing the so-called excitatory postsynaptic potential (EPSP). The regulatory effect of the GABAergic (mainly $GABA_A$ type) synapses is to repolarize V_m by the so-called inhibitory postsynaptic current (IPSC) producing the opposite effect on the membrane voltage and generating the inhibitory postsynaptic potential (IPSP).

The integration of the activity of these large amounts of inputs at the soma of the neuron determines the spiking behavior of the neuron (coding). Considering that on the average a neuron makes a single synapse to another neuron, each neuron receives contacts from thousands of neurons each of which try to send the information it carries. However, several neurons of a given area (sending area) can give each a single contact to the same neuron (receiving neuron). If the many neurons of the sending areas are excited by a stimulus, the integration of the synaptic responses on the receiving neuron will produce the postsynaptic representation of the stimulus. For example, several neurons of the dentate gyrus can input on the same neuron of one of the subfields of hippocampus (CA1 or CA3) [7]. Moreover, such an input interacts with the inputs coming from other neurons located on different areas (e.g., from the entorhinal cortex in the example given before [7]).

If we only look to this short and incomplete representation of the problem of the information management by a single neuron, it becomes clear how and why the correspondence between the sequence of spikes and the code it generates is very variable such that the correlation between the inputs and the code generated is unpredictable and appears random.

In this framework, a great amount of complexity depends on the mechanisms which regulate the transmission of a single bit (spike) information to each synapse. The synaptic activity is greatly influenced by many factors [8]. First of all, the way glutamatergic synapse contributes to the postsynaptic neuronal code depends strongly on the biophysical properties of the dendrite where the synapse is located and on the path from its location to the soma. The electrical signal generated at a synapse attenuates with distance according to the cable properties of the dendritic path which changes along the arborization depending mainly on the dendritic size [9, 10]. The attenuation with distance is of exponential type [9, 10]. Usually, the higher input impedance of the branches more far from the soma seems to help the diffusion of the far signals producing EPSP with higher amplitude, a phenomenon which some authors consider as a sort of "synaptic democracy" [11, 12].

To give an idea of some of the basic mechanisms involved in the modulation of the synaptic information transferred to a neuron, in the following we will briefly remember, in a simplified way, the basic mechanism of the synaptic transmission with a particular attention to those processes which participate to the modulation of the signal. A part of the ability to transmit and modulate the information depends directly on the synaptic structure. For this reason we will first describe a general glutamatergic synapses and after the pre- and postsynaptic mechanisms influencing the modulation of the information carried by a single bit of synaptic information (the EPSP). The modulatory effect on a sequence of elementary bits (a "word") will also be considered, and a final discussion will summarize the effects of the different modulatory systems.

2. Synaptic structure and mechanisms

A classical glutamatergic synapse is located on the top of a spine of the dendritic tree. The spine is composed of a neck, protruding from the dendritic shaft, and a head where the information is really received. A general description of the glutamatergic synapse includes a presynaptic button facing through a cleft with postsynaptic spine. The area of the presynaptic button, opposed to the postsynaptic spine, contains vesicles filled of glutamate and is called the active zone (AZ). A number of vesicle ranging 10–20 are anchored to the presynaptic membrane by the SNARE complex (soluble NSF attachment proteins) which is a protein complex docking the vesicles ready to be released [13]. The arrival of a presynaptic spike activates the fusion and the pore formation of a vesicle activated by the SNARE complex following the Ca^{2+} influx (see, e.g., [13]). The first step of the transfer of the single elementary bit of information is then the release of a vesicle of glutamate regulated by the SNARE complex following the arrival of a presynaptic spike. If we consider the spike as the elementary bit of the neuronal information carried, then we can consider the EPSP as the elementary bit of the synaptically coded information.

The presynaptic surface, containing the docked vesicles, is separated from the postsynaptic one by a distance (cleft) of ~20 nm. The synaptic cleft is a volume where the molecules of glutamate, released by the presynaptic vesicle, diffuse by Brownian motion [14]. The arrival of the presynaptic spike, thanks to the Ca^{2+} and the SNARE complex, induces the formation of a pore between a vesicle and the presynaptic membrane. This pore is the path followed by the glutamate molecules to transit from the vesicle to the synaptic cleft.

If we assume a generic horizontal section, the diameter of a cortical or hippocampal glutamatergic synapse ranges 0.2–1 μm [15–18]. Assuming an AZ of circular space and the cleft of ~20 nm, we get a volume of cylindrical space which many authors use to study the synaptic transmission by a computer modeling approach [14, 16, 19]. Not all the synaptic "cylinder" is free for the diffusion of glutamate. The AZ covers only a part of the whole synapse (mean radius 0.11 μm), while the surrounding part is occupied by fibrils which anchor the pre- and postsynaptic neuron [20–22].

At the postsynaptic side, two types of glutamate receptors are colocalized in an area which is almost of the same size of the AZ and is considered as of circular shape too (lower part of the cylinder) [22, 23]. This area is called postsynaptic density (PSD) and contains α-amino-3-hydroxy-5-methyl-4-isoxazolepropionic acid sensitive receptors (AMPA receptors) and N-methyl-d-aspartate sensitive receptors (NMDA receptors) which contains sites to bind glutamate molecules (but also for the glycine which is necessary for the synaptic activity). These two types of receptors have different roles in the transfer of synaptic information which we will discuss later in details. Both types are tetramers composed of a dimer of dimers [24]. As we will see later, the dimeric composition of the receptor plays an important role in shaping the postsynaptic response.

Apparently, the information transfer process is very simple in principle. The arrival of a presynaptic spike produces the fusion of a vesicle with the release of glutamate which activate postsynaptic receptors producing a depolarizing current (EPSC) which causes a variation of the postsynaptic membrane potential called EPSP which, diffusing through the dendritic branches, contribute at the soma, to the generation of the postsynaptic spike. However, any of the passage from the presynaptic to the postsynaptic side undergoes to a series of rearrangement of the information which makes the whole process extremely complex both to study and to interpret. In **Figure 1** a schematic representation of the information flow by synaptic transmission is presented.

Essentially, the different modulation systems produce a sort of complex nonlinear variability of the postsynaptic response. Variability of the EPSP is caused

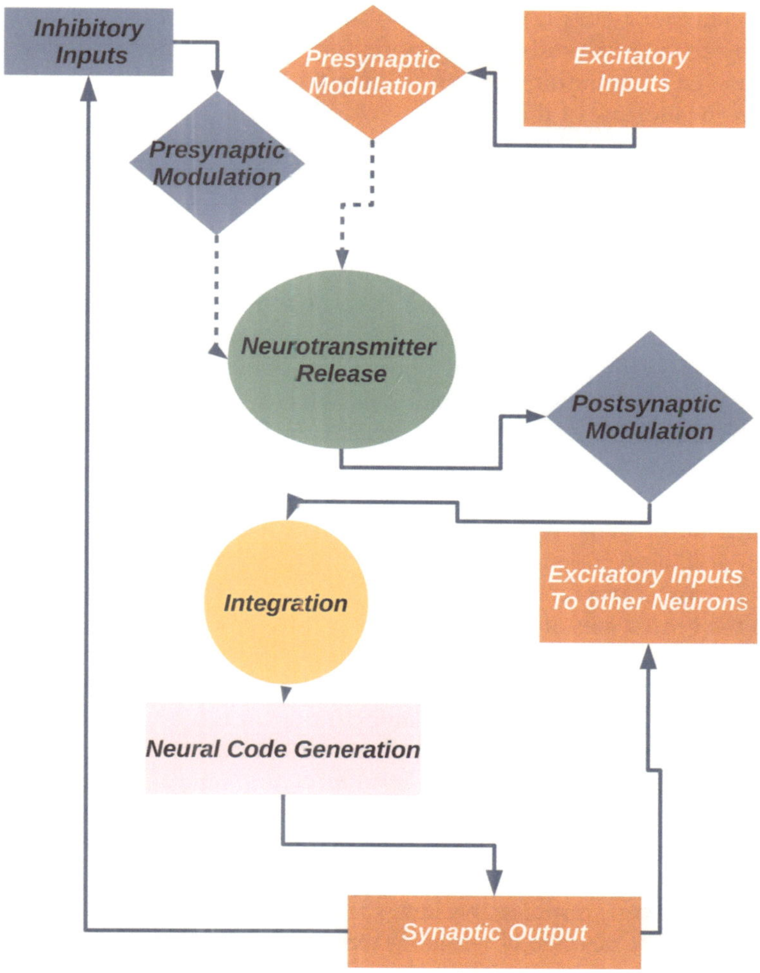

Figure 1.
Information flow by synaptic transmission.

both by pre- and postsynaptic mechanisms of control, and a part of this variability seems to be of stochastic nature (for a review see [8]). In the following we will examine some (but not all) regulatory mechanisms which imply modulation of variability of the postsynaptic response and their possible stochastic and/or deterministic nature. Some of the processes which produce synaptic response variability are external to the synapse and will be considered in an appropriate section.

2.1 Intrasynaptic factors of the EPSP variability

For intrasynaptic factors we mean those mechanisms operating at the level of the AZ or to the PSD area. For extrasynaptic we mean any influencer located or operating out of the synaptic "cylinder." Intrasynaptic factor can be divided into pre- and postsynaptic factors influencing the EPSP variability.

2.1.1 Presynaptic-dependent EPSP variability

A first point to stress about the presynaptic source of variability is that the probability of release of a vesicle following a presynaptic spike is not 1 and the range can be as wide as 0.2–0.91 [25, 26]. Modification of the releasing probability has been associated to the so-called presynaptic (non NMDA-dependent) long-term potentiation (LTP) assuming that the efficacy activity dependent of a synapse depends on the increase of the releasing probability [17, 27–29]. This point is crucial for the understanding on how the presynaptic neural code is coded synaptically because it means that not all the presynaptic spikes are coded by an EPSP. Moreover, if this probability changes as a function of the activity, this means that a different number of EPSP code for a given number of presynaptic spikes depending on the preceding activity. In terms of information, not all the presynaptic bits are transferred but only a fraction of it, and the size of the fraction is activity dependent. Moreover, the sequence of EPSPs does not sum linearly at the postsynaptic side [30, 31]. This means not only that only a part of bits composing the presynaptic "word" is transferred but also that their postsynaptic representation is extremely variable and depending on how many bits are transferred (which change as the probability of release changes with activity) and on the timing between the transferred bits.

Although the vesicular release is considered of quantal type, the release of single vesicle can produce different responses depending on several presynaptic factors (see, [2]). An important factor is the position of the vesicle (eccentricity) with respect to the central axis of the cylinder limited by the AZ and PSD. For a given configuration of the PSD (see next section), the release of glutamate from a more peripheral vesicle will produce an EPSC with smaller amplitude than one centered to AZ-PSD central axis [32–34]. Another important factor is the amount of molecules into the vesicle. Vesicle concentration, in fact, is extremely variable ranging 60–210 mM [15, 16, 35] with an average of ~140 mM. Assuming an internal radius of the vesicle with an average of 23 nm, it is clear that the number of molecules of glutamate released for a single bit of information is extremely variable. A variable number of molecules produce EPSC with different amplitude [32–34, 36–38]. In our early work on single glutamatergic response, we have considered the combination of the number of molecules and the position of release as stochastic factors [8, 32–34]. However, by considering the large variability of the concentration of glutamate in the vesicles, the thousand possible combinations of "position-number of molecules," this could be a powerful system of presynaptic regulation of the information transfer. In this respect, an interesting question arise: "what is the mechanism which, for a given presynaptic spike, 'decide' the correct combination 'position-number of Glutamate molecules'?" The SNARE complex, because of its different configurations depending on the membrane activity, could be a candidate for this decision role [13, 39, 40].

Although in the larger part of the cases a single vesicle opens with probability less than 1 for the arrival of a single presynaptic spike, in some cases a multivesicular release has been observed (see, e.g., [41, 42]). The multivesicular release found in some experiments opens many other interesting questions. The most relevant is: *what is the relationship on the number of vesicle opened for a spike the of information transferred?* Another interesting question is *What is the role of multivesicle release if usually a single release does not achieve postsynaptic saturation of the response?* [35, 43]. To summarize, the most important presynaptic factor of EPSP variability are:

- Probability of release of a vesicle following a single presynaptic spike and its dependence on the past activity
- Probability of multivesicular release
- Number of molecules inside the released vesicle

- Position (eccentricity) of the released vesicle
- Ca^{2+} different modulations of the SNARE complex

2.1.2 Postsynaptic-dependent EPSP variability

The PSD is the postsynaptic area containing the receptors (AMPA and NMDA) with different roles [44]. The number of receptors among different synapses is highly variable but varies also in the same synapse as a function of its maturity and activity. The number of AMPA receptors in a typical hippocampal synapse can range between 0 (AMPA-silent synapses [45]) and 80–100 [46, 47]. The number of AMPA is related to the synaptic maturation and potentiation and is strongly associated to phenomena like memory formation and learning (among many others [17, 29, 48–51]). Some of these mechanisms can change the properties of a synapse in the time-lapse of less than a second if the presynaptic neuron furnishes an appropriate stimulation. The variability of the number of AMPA not only produces potentiation of a synapse but also a depotentiation (by removing of AMPA [52]), and both mechanisms are Ca^{2+} and NMDA dependent [53]. AMPA can either be inserted (or removed) because a migration from the extrasynaptic membrane space to the PSD or just aquired from the cytoplasm [52]. According to some authors, also the number of NMDA receptors can change as a function of the activity [54]. This point is not trivial for the understanding of the synaptic response variability. By changing the number of receptors, it changes the total conductance and the current that the synapse can produce for a single presynaptic spike (see **Figure 2**).

Both AMPA and NMDA are tetramers (composed of four subunits) arranged as dimer of dimers. The dimeric and tetrameric composition produces a mosaic of configurations each with electrophysiological properties different from the others [24, 55–60]. Their conductances mediated over different dimeric compositions (as computed in Di Maio et al. Table 2 of [61]), in fact, are for AMPA 15 ± 10 pS and for NMDA 40 ± 15 pS. This means that the variability induced by the insertion activity-dependent of an AMPA, for example, will furnish a variation of the response depending on the dimeric composition (conductance) of the newly inserted receptor.

The current produced by the opening of the receptors (EPSC) produces a variation of the membrane potential (EPSP) at the postsynaptic side which depends on the biophysical properties of the postsynaptic membrane. The glutamatergic

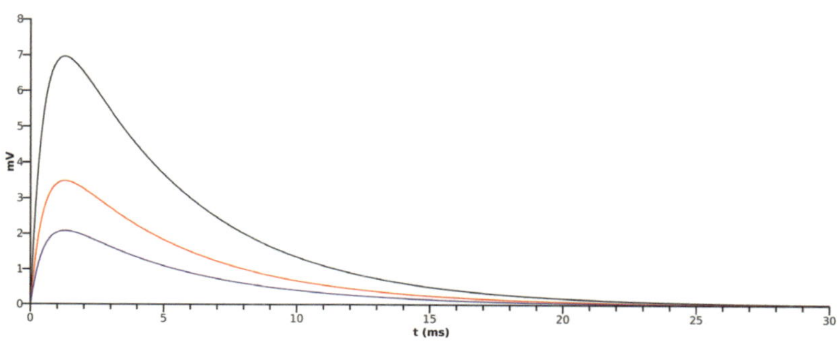

Figure 2.
Different synaptic responses obtained for the release of a single vesicle. The different amplitudes can be due either for presynaptic regulation (e.g., different positions of the vesicle or different numbers of molecules) or for postsynaptic regulation (e.g., different numbers of receptors or different membrane voltage at the moment of the EPSP start).

synapse is positioned at the top of a spine which is considered by many authors as a separate electrical compartment with a high input impedance [61–67]. The general equation which produces the EPSC is derived by Ohm's law:

$$I_{syn}(t) = g_{syn}(t)\left(V_m(t) - E_{syn}\right) \quad (1)$$

where I_{syn} is the current (EPSC) produced by AMPA and NMDA receptors, g_{syn} is the synaptic conductance, V_m is the membrane potential, and is the equilibrium potential computed by the Nernst equation considering all the ions (usually Na^+, K^+, and Ca^{2+}) involved in the synaptic current. The values of EPSP expressed as a variation of V_m depend on the input resistance (R_i) of the system:

$$V_m = I_{syn}R_i \quad (2)$$

As clear from Ohm's law, for a given current produced by the receptors, the variation of the membrane voltage amplitude depends on the value of . The spine in general is considered as a system with high input impedance (order of $G\Omega s$) [9, 10]. However, more recent papers have stressed that the spine circuit is rather complex and can be sub-compartmentalized [46, 62, 63, 67, 68]. However, two main parts of the spine compartmentalization play really a relevant role in shaping the postsynaptic response: the PSD area and the neck resistance [65, 66]. The neck resistance is the natural pathway for the signal to reach first the dendrite and then the soma and will be treated in the next section. About the PSD area, it is the area where the receptors are localized, and its characteristics directly influence the response of each single receptor. Being crowded of proteins, the resistive component of this is high, while the capacitive one is negligible [61, 64–66]. According to Eqs. (1) and (2), the current in this area can produce high variation of potential even for very small currents produced by the receptors (see the dependence of the EPSC and EPSP in **Figure 3** of [61]). PSD input resistance then is a key player in modulating the receptor current.

This is even more important if we consider the characteristics of the NMDA receptors and their contribution to the EPSC generation. At the resting level of the membrane potential ($V_r \sim -65$ mV), these receptors are blocked by Mg^{2+}, and, consequently, even if glutamate is release, they do not furnish a contribution to the EPSC. Mg^{2+}-block of NMDA is voltage dependent [55, 69]. The probability of NMDA receptor to give a contribution to the total conductance follows a sigmoid rule function depending on the membrane voltage. The complete unblocking of the total NMDA conductance (unblocking probability = 1) is obtained only for a very depolarized value of V_m ($V_m \sim +40$ mV) which is not a value in the usual range of action of the dendritic synapses [55, 61, 64–66, 69]. However, as we have shown in our recent works [61, 64] because of the PSD high input impedance, the current produced by the fast AMPA receptor activation can increase their probability to unblock and to contribute to the total synaptic conductance. It follows that different number and proportion of AMPA and NMDA produces different effects on the single EPSPs. This is an example of intrasynaptic receptor-dependent modulation of the EPSP which can be considered as due to the receptors' cooperativity [61].

The influence of the NMDA component on the total EPSP, being voltage dependent, does not only depends on the fast AMPA activation but also on external factors (see the following sections). In summary we can say that the postsynaptic processes involved in the variability of the EPSP are:

- The total number of postsynaptic receptors
- The relative number of postsynaptic receptors (AMPA versus NMDA)

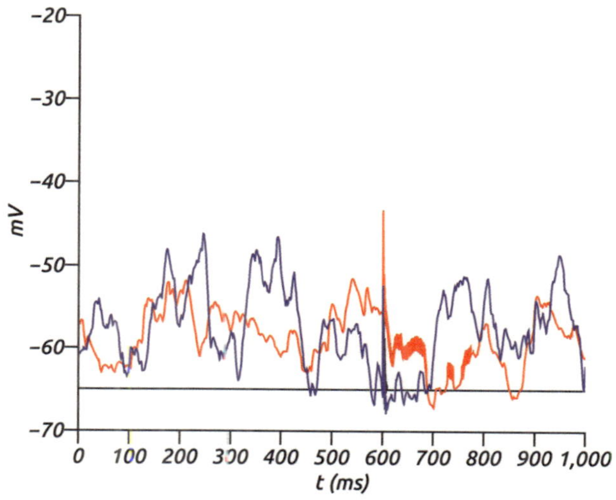

Figure 3.
Simulation of EPSP during dendritic activity produced by different firing frequencies of excitatory and inhibitory synapses. Depending on the firing frequencies of the other synapses located in the proximity, the membrane voltage under the spine has different amplitude and types of oscillation. The phase and the level of oscillation at the moment of the EPSP start, modulate the amplitude its amplitude and time course. In these simulations the EPSP occurred always at 600 ms. The black line is the time course of the membrane potential if no dendritic activity is present, and hence it is constant at the resting level (−65 mV).

- The diversity of the receptor conductance depending on the dimeric composition

- The biophysical properties of the PSD

2.2 Extrasynaptic factors of the EPSP variability

By looking just outside the restricted synaptic space, several other factors can influence the EPSP formation. In short we can say that, according to Eqs. (1) and (2), any factor which can influence the membrane potential in the proximity of the synapse can play a role in shaping the EPSP. The first important structure that we found outside of the synaptic space is the neck of the spine. It is the communication way between the synapse and the dendrite, and its electrical resistance determines the amount of information passed to the cell (dendrite). The value of the neck resistance is, then, crucial for the flow of information among different areas of the dendrites and the soma. Spine morphology is variable, and consequently its bio-electric properties [62, 67, 68, 70] and the presence of voltage-gated channels can further influence its ability to transfer the synaptic information [70]. According to some authors, the neck diameter and resistance are modulated also during a single synaptic event [62, 63, 67]. Modulation of the neck resistance produces, as a consequence, a modulation of the EPSP transmitted to the soma. However, the neck does not only carry the synaptic information to the dendrite. It acts also in the opposite direction by carrying the information on the state of dendrite to the PSD. In other words, the PSD is kept informed of the information arriving from other synapses located in the proximity. Dendritic activity, in fact, producing a difference of potential between the dendrite and the head of the spine, produces a net current, the direction of which depends on the difference of potential between the two structures. The current arriving from the dendrite is essentially amplified by the high input impedance of the PSD influencing strongly the total EPSP and the recruitment of the NMDA receptors [65, 66]. The neck of the spine is, then, a powerful modulator of the synaptic information transfer depending on the excitation (depolarization) level of the membrane potential in the dendrites [65, 66].

By considering the huge number of inputs (from 3×10^3 to 3×10^4) received, the dendritic arborization is not an electrical isopotential compartment. Differences of potential among the different branches can be due to different input activities and by spike backpropagation [71] in those areas where it can occur. The spike back propagation depends on the presence of Na^+ and/or Ca^{2+} voltage-gated channels in the dendrites [12, 63, 70, 72]. The presence and density of these channels differ both among neurons and, in the same neuron, among different dendritic regions [12, 63, 70, 72], and consequently differences of potential can produce complex potential waves transiting the dendrites, and this wave can reach the PSD trough the neck resistance influencing the single synaptic event. In our recent papers, we have studied a possible effect of potential waves produced by excitatory synaptic activity on the single synaptic response independently of the spike backpropagation [65, 66]. We have found that, depending on the number of active synapses and on their mean firing frequency, the amplitude, peak level, and time to peak of the response vary in a complex nonlinear fashion (see **Figure 3**) [65, 66]. The number of active synapses in some way simulates the input, for example, received from one area of the brain where a group of active neurons fire in a more or less synchronous way (in response to a stimulus) on the same neuron in a restricted dendritic area. For the case already mentioned, for example, a neuron of a hippocampal subfield can receive synchronous inputs from a large area (many neurons) of the dentate gyrus but also from areas of the Entorhinal cortex in separate regions of the dendritic branches (see, e.g., [7]). The neurons from one of these two areas fire with a mean frequency and a standard deviation which depends on the degree of their synchronization. Such a condition produces waves into the dendritic area interested to the stimulation which directly influence any single synapse which is active in the same time window [65, 66]. The membrane potential of the receiving neuron oscillates between two levels forming a voltage "band." The amplitude of this voltage "band" depends on the number of active synapses and on their mean firing frequency [65, 66]. The EPSP of a given synapse can occur at any level inside the "band." According to Eqs. (1) and (2), depending on the level at which the EPSP of a given synapse starts, its properties (amplitude, peak level, NMDA contribution, etc.) will change [65, 66]. In this band it is possible to identify a mean value which can be considered as the maximal likelihood level of V_m at which the EPSP can occur. This mean level increases (more depolarized) by increasing the number of active synapses and/or their firing frequency [65, 66]. The existence of this "band" of voltage furnish a large gamma of possible levels of V_m at which EPSP can occur and consequently it represents a very powerful regulator of the single EPSP depending on the time of occurrence (phase of the oscillation inside the band) [65, 66]. Said in a different way, the coincidence of the EPSP with the particular level of determines the type and amount of information the synapse transfers. NMDA receptors, being dependent on the membrane voltage for their activity, are especially sensitive to this kind of regulation, and in fact, the "coincidence" of the EPSP with the activity of other synapses is considered crucial for phenomena like LTP and memory which are NMDA dependent. These are the basic mechanisms who suggest that neurons, mostly in producing LTP and memory phenomena, act as *coincidence detectors* (among many others, see, e.g., [73]). The dendritic activity modulatory effect on the transfer of a single bit of synaptic information depends essentially on the variation of potential in the membrane and can be summarized as due to:

- Spike backpropagation
- Active synaptic inputs on the dendritic tree

3. Discussion

This short overview was aimed to stress how the information transferred among synapses and its elaboration undergo to many regulation systems which involve structural, functional, and cooperative processes. By identifying the EPSP as the elementary bit of the information transferred by a single synapse, we have outlined some of the pre- and postsynaptic sources of variability.

In general the word "variability" can be used with two meanings. It can be attributed either to something which vary in an unpredictable way, or it can mean the possibility to change following specific actions. This is especially true for the causes of variability of the EPSP. EPSP variability can be due (a) to stochastic processes [8, 32–34] or (b) to specific systems of regulation which operate at different levels of the synaptic transmission (intrasynaptic or extrasynaptic). About the stochastic variability, we cannot say too much. If a process occurs randomly, we can only try to understand its effects observing the responses and trying to explain the phenomena by a plausible model which (statistically) describes the natural event. The big problem in this respect is to identify if this type of system depends really on stochastic processes or if stochasticity is apparent because the lack of the full information needed to characterize the processes. From the most top point of view, almost all the causes of EPSP variability described herein can appear of stochastic type [8, 32–34], but we cannot definitively exclude that the apparent stochasticity is due to our incomplete understanding of all the underlying mechanisms and/or to the lack of knowledge of all the steps underlying the process. Just to give an example, if we consider the response variability depending on the number of molecules in the vesicle, its position on the AZ (eccentricity), and its variable release probability, [8, 17, 27–29, 32–34] we can assume a stochastic origin of the presynaptic factors of the synaptic response variability. However, the mechanism of the vesicle opening is under the control of the SNARE complex which is intimately connected to the vesicle and is the responsible for the Ca^{2+}dependent pore opening. This complex can have different configurations depending on the state of the neuron (see, e.g., [13]). We cannot exclude that a more complete understanding of the SNARE complex functionality could permit the definition of a relationship between the information passed by the synapse and the characteristics (position and number of molecules) of the released vesicle. This is only a possibility for one of the many regulatory factors involved in the synaptic response modulation, and their discussion is not in the goal of the present chapter. The important point that we want to stress is to outline the large variability of the EPSP and that this variability is controlled by many different systems. Variability, then, in the context of this chapter, has to be intended as the ability to be modulated ("tuning") of the system.

The tuning of the information to transfer is not only due to the pre- and postsynaptic neuron. The activity of thousands of synapses inputting on a neuron produces waves of potential into the dendritic tree which directly influence the characteristics of the information transferred by each single synapse [65, 66]. Even two single synapses, closely located on a dendritic branch, influence each other. The synapse which fires first, in fact, by changing the membrane potential influences the response of the synapse firing later if the time interval between the two events is compatible with the decay time of the first event [30].

Two key questions emerge by the above considerations: (a) *what is the characteristic of response (EPSP or EPSC) which better represent the code of the single bit of the synaptic information?* and (b) *how does the single bit of synaptic information produces a synaptic code at the postsynaptic level?* The two questions are not independent to each other. The best candidate to code for the single bit of information seems to be the EPSP (or EPSC) "amplitude." The amplitude depends on the characteristic of the synapse (number of receptors, PSD input impedance, spine neck resistance, etc.) and on the activity of the dendrite on which the synapse is hosted.

The amplitude of an EPSP occurring when the postsynaptic membrane is close to the reverse potential 0 mV can approach 0 mV. This means that the postsynaptic mechanism of tuning can, depending on its state, nullify the information. Alternatively, an EPSP starting when the postsynaptic membrane potential is close to the resting potential (or even in a hyperpolarized state), the amplitude is maximized [65, 66].

Interestingly, if the single bit of information is coded by the EPSP (EPSC) amplitude, while a diffuse excitation depolarizing the membrane reduces the amount of information passed by the synapse, the inhibition works in the opposite direction. Driving the membrane potential far from the reverse potential, in fact, the inhibitory inputs play a favor of increasing the amplitude [30].

Assuming that the single spike represents the single bit of information of a neuron, a sequence of spikes emitted by a presynaptic neuron represent a "word" that is the full representation of a stimulus in that neuron. The synaptic codification of this "word" should be an equivalent sequence of EPSP. This does not always hold. As we have said, the probability that an EPSP is generated when a spike arrives is less than 1. Moreover, EPSPs sum non linearly at the postsynaptic side and the amplitude and shape of the resulting sum depend on the time between the EPSPs. In addition, the different EPSPs are modulated postsynaptically each differently depending on the coincidence of their start and the phase of the wave produced by the dendritic activity. The same presynaptic "word" can then have different postsynaptic representations since formed by different number of EPSPs coded with different amplitudes and presenting different shapes and duration because of the different NMDA contributions. In short, rarely the same repeated stimulus represented by a sequence of spikes will have a fixed clearly identifiable representation at the postsynaptic side. This variability of synaptic representation of a "word" is probably the main cause of the variability of the postsynaptic neuronal "word" (different sequences of the postsynaptic spikes). This means that the single presynaptic "word" almost never determines the postsynaptic spike sequence (postsynaptic "word") which is always the results of the cooperation of all the inputs arriving in a given time window. Although in many experimental results it is possible to identify a sort of relationship between a stimulus and some characteristics of the spikes sequence it induces in a given neuron, probably in the real brain, the situation is much more complex.

A last comment on how the mechanisms of postsynaptic regulation play a role in the information processing by considering the different information arriving from many neurons on a single one. If we consider the inputs on a single neuron coming from two areas of the brain and located in the close proximity on the dendritic tree, the area which sends early the information can inhibit the information of the other area. A massive input arriving from many excited neurons of a firstly activated area will produce a strong depolarization of the dendritic area which will inhibit (if not nullify) the information arriving from the other area. This can be probably a mechanism which regulates, at the single neuron level, the competition between two antagonist inputs involving different areas of the brain but also a mechanism of "decision-making." The priority for the response, in this case, is time dependent since the single neuron will furnish mostly a response to the area activated early. However, the inhibitory regulation of these mechanisms can produce several different levels of single-neuron response to the two different stimuli. In other words, the mechanisms of regulation of the synaptic information transfer based on the variation of the membrane potential regulate also the competition and/or the level of integration of the information arriving from different areas of the brain on close areas of the dendritic tree of the same neuron.

A last comment on the nature of the codification of the synaptic information and on the computational ability of the dendrites is necessary. While one can discuss on the digital or analogical nature of the neural code which is based on stereotyped spike (bit) sequence, the same does not hold for the transformation of the neural code into the synaptic code at the dendritic level. EPSPs are not stereotyped (all or none) systems, and, as shown before, their representation of the presynaptic "words" change the number of bits, shape, and amplitude. By looking this type of synaptic codification, we would exclude a "dendritic computation" based on algebraic-like or Boolean-type computation (see, e.g., [31]). Most probably, dendritic computation has to be a sort of analog computation which still remains to be understood.

4. Conclusion

In this chapter we have given a *non-exhaustive light* overview on how the synaptic response is modulated by several intrinsic and extrinsic factors acting at different stages of the process of the information processing and transfer among neurons.

A first important point that should emerge from what exposed is that the problem, also at the level of the single synapse, is extremely complicated by the different effects produced by the many systems of modulation of the information.

A second, but not less, important point is that our knowledge of the information transfer by synaptic transmission is still very poor although a great effort is spent in this direction.

The different levels, at which the regulation of the information processing mediated by synapse occurs, require the cooperativeness of different scientific approaches. The experimental methodologies and paradigms of investigations, although improving day by day, cannot answer alone all the questions still open because of the experimental technique limitations. A good synergy between experimental, theoretical, and computational modeling approaches is needed. The possibility to use big computational facility becomes a limiting factor for the success.

The unveiling of the synaptic mechanisms of information processing and transfer is of great importance because information processing is the key ability of the living systems to survive in the environment and, for the humans, is also the key ability for high-level cognitive performance. As stressed in the introduction, the loss of cognitive performance, like in the Alzheimer and in the Parkinson diseases, is strongly associated to the synaptic malfunctioning. Memory and learning are essentially synaptic functions.

In addition, the investigation on synaptic information processing and in the synaptic functionality also support the researches in other fields as, for example, in projecting and realizing artificial computational systems which, by using the powerful mechanism of synaptic information processing, tray to produce high-performance artificial system (see, e.g., [74]).

Some important challenges for the future studies of the information processing mediated by synapses can be summarized as follows:

- Unveiling the presynaptic mechanism involved in vesicle release as function of the presynaptic spike sequence.
- Decoding of the synaptic EPSP sequence as function of the presynaptic spike sequence and its relationship with the presynaptic stimuli which determine the sequence.
- Decoding the real integration systems which relate the information arriving from different areas and their integration at the dendritic level in order to produce the postsynaptic spike sequence.
- Establishing the relationship between the dendritic excitation produced by thousands of synapses and the single synaptic event. This is necessary to understand the real contribution of a single event and of a single presynaptic sequence of spikes in building the postsynaptic EPSP sequence and consequently its participation to the postsynaptic spike sequences (postsynaptic neural code).

All these challenges are very hard, and each of them will need still years of investigation in the field of the information processing in the brain.

Author details

Vito Di Maio* and Silvia Santillo
Institute of Applied Science and Intelligent Systems (ISASI) of CNR, Pozzuoli (NA), Italy

*Address all correspondence to: vito.dimaio@cnr.it

References

[1] Gardoni F, Di Luca M. Targeting glutamatergic synapses in Parkinson's disease. Current Opinion in Pharmacology. 2015;**20**:24-28. DOI: 10.1016/j.coph.2014.10.011

[2] Sheng M, Sabatini BL, Südhof TC. Synapses and Alzheimer's disease. Cold Spring Harbor Perspectives in Biology. 2012;**4**:1-18. DOI: 10.1101/cshperspect.a005777

[3] Rojas DC. The role of glutamate and its receptors in autism and the use of glutamate receptor antagonists in treatment. Journal of Neural Transmission. 2014;**121**:891-905. DOI: 10.1007/s00702-014-1216-0

[4] Gulyás AI, Megías M, Emri Z, Freund TF. Total number and ratio of excitatory and inhibitory synapses converging onto single interneurons of different types in the CA1 area of the rat hippocampus. Journal Neuscience. 1999; **19**:10082-10097

[5] Megías M, Emri Z, Freund TF, Gulyás AI. Total number and distribution of inhibitory and excitatory synapses on hippocampal CA1 pyramidal cells. Neuroscience. 2001; **102**:527-540

[6] Villa KL, Nedivi E. Excitatory and inhibitory synaptic placement and functional implications. In: Emoto K, Wong R, Huang E, Hoogenraad C, editors. Dendrites. Japan: Springer; 2016. pp. 467-487. DOI: 10.1007/978-4-431-56050-0-18

[7] Bartesaghi R, Di Maio V, Gessi T. Topographic activation of the medial entorhinal cortex by presubicular commissural projections. Journal of Comparative Neurology;**487**(3): 283-299. DOI: 10.1002/cne.20547. Available from: https://onlinelibrary.wiley.com/doi/abs/10.1002/cne.20547

[8] Di Maio V, Ventriglia F, Santillo S. Stochastic, structural and functional factors influencing AMPA and NMDA synaptic response variability: A review. Neuronal Signaling. 2017;**1**:1-11. DOI: 10.1042/NS20160051

[9] Rall W. Electrophysiology of a dendritic neuron model. Biophysical Journal. 1962;**2**:145-167

[10] Rall W, Rinzel J. Branch input resistance and steady attenuation for input to one branch of a dendritic neuron model. Biophysical Journal. 1973;**13**:648-688

[11] Häusser M. Synaptic function: Dendritic democracy. Current Biology. 2001;**11**(1):R10-R12. DOI: 10.1016/S0960-9822(00)00034-8. ISSN 0960–9822. Available from: http://www.sciencedirect.com/science/article/pii/S0960982200000348

[12] Rumsey CC, Abbott LF. Synaptic democracy in active dendrites. Journal of Neurophysiology. 2006;**96**(5): 2307-2318. DOI: 10.1152/jn.00149.2006

[13] Han J, Pluhackova K, Böckmann RA. The multifaceted role of snare proteins in membrane fusion. Frontiers in Physiology. 2017;**8**:5. DOI: 10.3389/fphys.2017.00005. Available from: https://www.frontiersin.org/article/10.3389/fphys.2017.00005

[14] Ventriglia F, Di Maio V. A Brownian simulation model of glutamate synaptic diffusion in the femtosecond time scale. Biological Cybernetics. 2000;**83**: 93-109

[15] Clements JD, Lester RA, Tong G, Jahr CE, Westbrook GL. The time course of glutamate in the synaptic cleft. Science. 1992;**258**(5087):1498-1501. DOI: 10.1126/science.1359647. ISSN: 0036-8075

[16] Clements JD. Transmitter timecourse in the synaptic cleft: Its role in central synaptic function. Trends in Neurosciences. 1996;**19**:163-171

[17] Meldolesi J. Long-term potentiation. The cell biology connection. Current Biology. 1995;**5**(9):1006-1008

[18] Gu B-M, van Rijn H, Meck WH. Oscillatory multiplexing of neural population codes for interval timing and working memory. Neuroscience and Biobehavioral Reviews. 2015;**48**: 160-185. DOI: 10.1016/j.neubiorev.2014.10.008. Available from: http://www.sciencedirect.com/science/article/pii/S0149763414002589

[19] Agmon N, Edelstein AL. Collective binding properties of receptor arrays. Biophysical Journal. 1997;**72**:1582-1594

[20] Zuber B, Nikonenko I, Klauser P, Muller D, Dobochet J. The mammalian central nervous synaptic cleft contains a high density of periodically organized complexes. Proceedings of the National Academy of Sciences of the United States of America. 2005;**102**: 19192-19197

[21] Ventriglia F. Effect of filaments within the synaptic cleft on the response of excitatory synapses simulated by computer experiments. Biosystems. 2011;**104**:14-22

[22] Ventriglia F, Di Maio V. Effects of AMPARs trafficking and glutamate-receptor binding probability on stochastic variability of EPSC. Biosystems. 2013;**112**:298-304

[23] Ventriglia F, Di Maio V. Glutamate-AMPA interaction in a model of synaptic transmission. Brain Research. 2013;**1536**:168-176

[24] Tichelaar W, Safferling M, Keinänen K, Stark H, Madden DR. The three-dimensional structure of an ionotropic glutamate receptor reveals a dimer-of-dimers assembly. The Journal of Molecular Biology. 2004;**344**: 435-442. DOI: 10.1016/j.jmb.2004.09.048

[25] Dobrunz LE, Stevens CF. Heterogeneity of release probability, facilitation, and depletion at central synapses. Neuron. 1997;**18**:995-1008

[26] Park H, Li Y, Tsien RW. Influence of synaptic vesicle position on release probability and exocytotic fusion mode. Science. 2012;**335**(6074):1362-1366. DOI: 10.1126/science.1216937

[27] Kokaia M. Long-term potentiation of single subicular neurons in mice. Hippocampus. 2000;**10**(6):684-692. DOI: 10.1002/1098-1063(2000)10:6

[28] Zakharenko SS, Zablow L, Siegelbaum SA. Visualization of changes in presynaptic function during long-term synaptic plasticity. Nature Neuroscience. 2001;**4**(7):711-717. DOI: 10.1038/89498. ISSN: 1097-6256. Available from: http://tinyurl.sfx.mpg.de/qv9r; http://www.bibsonomy.org/bibtex/2b66228cbabc54500b35eb408436fdbd2/schrod

[29] Raymond CR. LTP forms 1, 2 and 3: Different mechanisms for the “long” in long-term potentiation. Trends in Neurosciences. 2007;**30**(4):167-175. DOI: 10.1016/j.tins.2007.01.007

[30] Di Maio V. Regulation of information passing by synaptic transmission: A short review. Brain Research. 2008;**1225**:26-38

[31] Hao J, Wang XD, Yang D, Poo MM, Zhang X. An arithmetic rule for spatial summation of excitatory and inhibitory inputs in pyramidal neurons. Proceedings of the National Academy of Sciences. 2009;**106**:21906-21911. DOI: 10.1073/pnas.0912022106. ISSN: 0027-8424. Available from: https://www.pnas.org/content/early/2009/12/01/0912022106

[32] Ventriglia F, Di Maio V. Stochastic fluctuation of the synaptic function. Biosystems. 2002;**67**:287-294

[33] Ventriglia F, Di Maio V. Synaptic fusion pore structure and AMPA receptors activation according to Brownian simulation of glutamate diffusion. Biological Cybernetics. 2003; **88**:201-209

[34] Ventriglia F, Di Maio V. Stochastic fluctuation of the quantal EPSC amplitude in computer simulated excitatory synapses of hippocampus. Biosystems. 2003;**71**:195-204

[35] Liu G, Choi S, Tsien RW. Variability of neurotransmitter concentration and nonsaturation of postsynaptic AMPA receptors at synapses in hippocampal cultures and slices. Neuron. 1999;**22**: 395-409

[36] Schikorski T, Stevens CF. Morphological correlates of functionally defined synaptic vesicle populations. Nature Neuroscience. 2001;**4**:391-395

[37] Karunanithi S, Marin L, Wong K, Atwood HL. Quantal size and variation determined by vesicle size in normal and mutant Drosophila glutamatergic synapses. The Journal of Neuroscience. 2002;**22**(23):10267-10276. ISSN: 1529-2401

[38] Richards DA. Vesicular release mode shapes the postsynaptic response at hippocampal synapses. Journal of Physiology. 2009;**587**(Pt 21):5073-5080. DOI: 10.1113/jphysiol.2009.175315

[39] Han X, Jackson MB. Structural transitions in the synaptic SNARE complex during Ca^{2+} -triggered exocytosis. The Journal of Cell Biology. 2006;**172**(2):281-293. DOI: 10.1083/ jcb.200510012

[40] Kiyonaka S, Wakamori M, Miki T, Uriu Y, Nonaka M, Bito H, et al. Rim1 confers sustained activity and neurotransmitter vesicle anchoring to presynaptic Ca^{2+} channels. Nature Neuroscience. 2007;**10**:691-701. DOI: 10.1038/nn1904

[41] Boucher J, Kröger H, Sík A. Realistic modelling of receptor activation in hippocampal excitatory synapses: Analysis of multivesicular release, release location, temperature and synaptic cross-talk. Brain Structure and Function. 2010;**215**:49-65

[42] Rudolph S, Tsai M-C, von Gersdorff H, Wadiche JI. The ubiquitous nature of multivesicular release. Trends in Neurosciences. 2015. doi: 10.1016/j. tins.2015.05.008;**38**(7):428-438. ISSN: 0166-2236. Available from: http://www. sciencedirect.com/science/article/pii/ S0166223615001228

[43] McAllister AK, Stevens CF. Nonsaturation of AMPA and NMDA receptors at hippocampal synapses. Proceedings of the National Academy of Sciences of the United States of America. 2000;**97**:6173-6178

[44] Nusser Z. AMPA and NMDA receptors: Similarities and differences in their synaptic distribution. Current Opinion in Neurobiology. 2000;**10**(3): 337-341. DOI: 10.1016/S0959-4388(00) 00086-6. Available from: http://www. sciencedirect.com/science/article/pii/ S0959438800000866

[45] Liao D, Scannevin RH, Huganir R. Activation of silent synapses by rapid activity-dependent synaptic recruitment of ampa receptors. Journal of Neuroscience. 2001;**21**(16): 6008-6017. DOI: 10.1523/ JNEUROSCI.21-16-06008.2001. Available from: http://www.jneurosci. org/content/21/16/6008

[46] Hill TC, Zito K. LTP-induced long-term stabilization of individual nascent dendritic spines. Journal of Neuroscience. 2013;**33**(2):678-686. DOI: 10.1523/ JNEUROSCI.1404-12.2013. Available

from: http://www.jneurosci.org/content/33/2/678

[47] Czöndör K, Thoumine O. Biophysical mechanisms regulating AMPA receptor accumulation at synapses. Brain Research Bulletin. 2013; **93**:57-68. DOI: 10.1016/j.brainresbull.2012.11.001

[48] Larkman AU, Jack JJ. Synaptic plasticity: Hippocampal LTP. Current Opinion in Neurobiology. 1995;**5**(3): 324-334

[49] Malinow R, Malenka RC. AMPA receptor trafficking and synaptic plasticity. The Annual Review of Neuroscience. 2002;**25**:103-126. DOI: 10.1146/annurev.neuro.25.112701.142758

[50] Nicoll R, Schmitz D. Synaptic plasticity at hippocampal mossy fibre synapses. Nature Reviews. Neuroscience. 2005;**6**:863-876

[51] Rao VR, Finkbeiner S. NMDA and AMPA receptors: Old channels, new tricks. Trends in Neurosciences. 2007; **30**(6):284-291. DOI: 10.1016/j.tins.2007.03.012

[52] Sanderson TM, Collingridge GL, Fitzjohn SM. Differential trafficking of AMPA receptors following activation of NMDA receptors and mGluRs. Molecular Brain. 2011;**4**(1):30. DOI: 10.1186/1756-6606-4-30. Available from: http://www.molecularbrain.com/content/4/1/30

[53] Bliss TVP, Collingridge GL. Expression of nmda receptor-dependent LTP in the hippocampus: Bridging the divide. Molecular Brain. 2013;**6**:1-14

[54] Watt AJ, Sjöström PJ, Häusser M, Nelson SB, Turrigiano GG. A proportional but slower nmda potentiation follows ampa potentiation in LTP. Nature Neuroscience. 2004;**7**: 518-524. DOI: 10.1038/nn1220

[55] Jahr CE, Stevens CF. Voltage dependence of NMDA-activated macroscopic conductances predicted by single-channel kinetics. The Journal of Neuroscience. 1990;**10**:3178-3182

[56] Dingledine R, Borges K, Bowie D, Traynelis SF. The glutamate receptor ion channels. Pharmacological Reviews. 1999;**51**:7-61

[57] Smith TC, Wang LY, Howe JR. Heterogeneous conductance levels of native AMPA receptors. Journal of Neuroscience. 2000;**20**(6):2073-2085

[58] Mayer ML. Glutamate receptor ion channels. Current Opinion in Neurobiology. 2005;**15**(3):282-288. DOI: 10.1016/j.conb.2005.05.004. ISSN: 0959-4388. Available from: http://www.sciencedirect.com/science/article/pii/S0959438805000693. Signalling mechanisms

[59] Greger IH, Ziff EB, Penn AC. Molecular determinants of AMPA receptor subunit assembly. Trends in Neurosciences. 2007;**30**(8):407-416. DOI: 10.1016/j.tins.2007.06.005

[60] Traynelis SF, Wollmuth LP, CJ MB, Menniti FS, Vance KM, Ogden K, et al. Glutamate receptor ion channels: Structure, regulation, and function. Pharmacological Reviews. 2010;**62**: 405-496

[61] Di Maio V, Ventriglia F, Santillo S. AMPA/NMDA cooperativity and integration during a single synaptic event. Journal of Computational Neuroscience. 2016;**41**:127-142

[62] Araya R, Jiang J, Eisenthal KB, Yuste R. The spine neck filters membrane potentials. Proceedings of the National Academy of Sciences of the United States of America. 2006;**103**(47):17961-17966. DOI: 10.1073/pnas.0608755103

[63] Palmer LM, Stuart GJ. Membrane potential changes in dendritic spines

during action potentials and synaptic input. Journal of Neuroscience. 2009; **29**(21):6897-6903. DOI: 10.1523/ JNEUROSCI.5847-08.2009. ISSN: 0270-6474. Available from: http://www. jneurosci.org/content/29/21/6897

[64] Di Maio V, Ventriglia F, Santillo S. A model of cooperative effect of AMPA and NMDA receptors in glutamatergic synapses. Cognitive Neurodynamics. 2016;**10**:315-325

[65] Di Maio V, Santillo S, Sorgente A, Vanacore P, Ventriglia F. Influence of active synaptic pools on the single synaptic event. Cognitive Neurodynamics. 2018. DOI: 10.1007/ s11571-018-9483-3. ISSN: 1871-4099. Available from: https://doi.org/10.1007/ s11571-018-9483-3

[66] Di Maio V, Santillo S, Ventriglia F. Multisynaptic cooperation shapes single glutamatergic synapse response. Brain Research. 2018;**1697**:93-104. DOI: 10.1016/j.brainres.2018.06.016. ISSN: 0006-8993. Available from: https:// www.sciencedirect.com/science/article/ pii/S0006899318303421

[67] Tønnesen J, Rózsa G, Katona B, Nägerl UV. Spine neck plasticity regulates compartmentalization of synapses. Nature Neuroscience. 2014;**17**: 678-685

[68] Weber JP, Andrásfalvy BK, Polito M, Magó Á, Ujfalussy BB, Makara JK. Location-dependent synaptic plasticity rules by dendritic spine cooperativity. Nature Communications. 2016;**7**:1-14. DOI: 10.1038/ncomms11380

[69] Vargas-Caballero MI, Robinson HP. Fast and slow voltage-dependent dynamics of magnesium block in the NMDA receptor: The asymmetric trapping block model. The Journal of Neuroscience. 2004;**24**:6171-6180

[70] Araya R, Nikolenko V, Eisenthal KB, Yuste R. Sodium channels amplify spine potentials. Proceedings of the National Academy of Sciences of the United States of America. 2007; **104**(30):12347-12352. DOI: 10.1073/ pnas.0705282104

[71] Rozsa B, Zelles T, Vizi ES, Lendvai B. Distance-dependent scaling of calcium transients evoked by backpropagating spikes and synaptic activity in dendrites of hippocampal interneurons. Journal of Neuroscience. 2004;**24**(3):661-670. DOI: 10.1523/ JNEUROSCI.3906-03.2004. ISSN: 0270-6474. Available from: http://www. jneurosci.org/content/24/3/661

[72] Nuriya M, Jiang J, Nemet B, Eisenthal KB, Yuste R. Imaging membrane potential in dendritic spines. Proceedings of the National Academy of Sciences of the United States of America. 2006;**103**(3):786-790. DOI: 10.1073/pnas.0510092103

[73] Tabone CJ, Ramaswami M. Is NMDA receptor- coincidence detection required for learning and memory? Neuron. 2012;**74**(5):767-769. DOI: 10.1016/j.neuron.2012.05.008

[74] Chen Y, Yu H, Gong J, Ma M, Han H, Wei H, et al. Artificial synapses based on nanomaterials. Nanotechnology. 2018;**30**(1):012001. DOI: 10.1088/ 1361-6528/aae470

Chapter 8

Computer Simulations of Hippocampal Mossy Fiber Cleft Zinc Movements

Johnattan C.S. Freitas, João N. Miraldo, Carlos Manuel M. Matias, Fernando D.S. Sampaio dos Aidos, Paulo J. Mendes, José C. Dionísio, Rosa M. Santos, Luís M. Rosário, Rosa M. Quinta-Ferreira and Emília Quinta-Ferreira

Abstract

Zinc ions have key regulatory, structural, and catalytic functions and mediate a variety of intra- and intercellular processes. The hippocampal mossy fiber boutons contain large amounts of free or loosely bound vesicular zinc, which can be co-released with glutamate. Zinc can interact with a variety of ionic channels (N-VDCCs, L-VDCCs, K_{ATP}), glutamate receptors (AMPA, KA, NMDA 2A, 2B), glutamate transporters (GLAST, EAAT4), and molecules (ATP). The dynamic properties of cleft free, complexed, and total zinc were addressed, considering the known concentration and affinity of various cleft zinc sensitive sites, mainly in the post-synaptic area and in glial cells. The computer model included three different zinc release processes, with short, medium, and long duration, described, like the uptake ones, by alpha functions. The results suggest that, depending on the amount of release, zinc clearance is largely due, either, to zinc binding to NMDA 2A receptor sites or to glial GLAST transporters.

Keywords: synaptic modeling, zinc-binding sites and complexes, glutamate receptors and transporters, zinc clearance and uptake, CA3 area

1. Introduction

Zinc is one of the most concentrated trace elements in the brain, being essential for normal cellular function and signaling processes in the central nervous system (CNS) [1–3]. This system contains very large amounts of chelatable or free zinc [4], mainly in the synaptic vesicles of excitatory nerve terminals [5], essentially in the hippocampal mossy fibers from CA3 area [1, 2]. After release, zinc affects the behavior of several voltage-gated and receptor-operated ionic channels [6–12]. The action of zinc in different types of receptors and channels depends essentially on two factors: their concentration in the synapses and their affinity for zinc. A clear understanding of the action of zinc in individual binding sites is restricted by the complexity of the synaptic transmission process. To further investigate this zinc role, a computational model was elaborated to describe zinc changes associated with the most important zinc-binding sites in the synaptic cleft between a mossy fiber terminal and a pyramidal cell of the hippocampal CA3 area, as previously reported [12]. Assuming that zinc is co-released with the neurotransmitter glutamate, this model was constructed taking into account previous studies that include computer simulations of glutamate dynamics in the synaptic cleft [13–16]. In the zinc model [12], the variation of total cleft zinc changes is obtained by subtracting two alpha functions, describing zinc release and zinc uptake. These functions, characterized by rapid climb phases and slower decays, were determined based on the assumed maximum amplitude and rising time values of cleft free zinc concentration. The corresponding parameters were defined taking into account experimental results, from optical and electrophysiological zinc experiments, reporting cleft zinc changes [17–27].

The process of zinc clearance from the synaptic cleft may include various actions, such as zinc binding, uptake, and entry into postsynaptic cells. Zinc can bind to a variety of pre- and/or postsynaptic receptors, voltage-dependent ionic channels, glutamate transporters expressed in glial cells, and also free molecules in the cleft medium. One of the most important targets for zinc action is the N-methyl-D-aspartate (NMDA) receptor-binding site with a high affinity for zinc [28]. However, it is present only at low concentrations in the mossy fiber terminals (about 80 nM) [29]. The most abundant zinc-binding site is the GLAST glial glutamate transporter, which is responsible for glutamate removal from the synaptic cleft into the glial cells [30, 31]. Zinc also forms complexes with 2-amino-3-(3-hydroxy-5-methyl-isoxazol-4-yl), propanoic acid (AMPA) and kainic acid (KA) glutamate receptors [29, 32–34], and potassium-ATP (K_{ATP}) channels [35, 36], with another type of glutamate transporter, the EAAT4 [37–39], and with the L- and N-types of voltage-dependent calcium channels (VDCCs) [40–42]. The concentration, affinity, and kinetics of zinc-binding sites are included in the model and have a very large impact on the behavior of zinc changes. On the other hand, the zinc uptake process is largely unknown, being probably mediated by zinc transporters and/or by zinc movements evoked by the electrochemical gradient [43–46]. In this model it was assumed that uptake is much slower than release, and a time constant was chosen for the latter process that is much larger than the time constant for the former (see Section 2). With respect to zinc entry into postsynaptic neurons, which may include NMDA receptors, voltage-dependent calcium channels, calcium-permeable AMPA/kainate channels, and the Na/Ca exchanger [17, 18, 32, 34, 47], it is considered that even the strongest stimulation protocol considered in this study is not strong enough to open the postsynaptic zinc permeant routes. In the present model, three different stimulation protocols were considered, named single (zinc release associated with a single stimulus), short, and long (multiple release processes that last for more times). It was considered that those stimulation processes evoked maximum cleft free zinc concentrations of 10 nM, 100 nM, and 1 μM, respectively. These concentrations are close to the values suggested in previous studies performed with similar types of stimulation [11, 19, 21, 27, 48]. Thus, in the present model, only cleft zinc concentrations below or equal to 1 μM were considered; therefore, no zinc enters the postsynaptic region, which may only occur for higher cleft zinc values [18, 20, 45].

The mossy fiber synapses have a very narrow synaptic cleft, measuring less than 20 nm [49–51]. For that reason, the movement of released glutamate and zinc, with similar free diffusion coefficients, is very rapid, reaching the opposite side of the cleft in a few microseconds. In hippocampal neurons, following an instantaneous release from a vesicle, the decay of glutamate concentration is very fast (tens of microseconds), being reduced to an almost constant value in about 50 μs [16]. Despite the more complex geometry of the mossy fibers [52], it can be assumed that the zinc concentration has a similar time course. For this reason, the diffusion was not included in the model, and it was considered that the cleft free zinc concentration is uniform during the binding process.

2. Methods

2.1 Model equations

The total amount of zinc in the synaptic cleft is, at any one time, partly bound to a number of sites and partly unbound. We shall refer to the concentration of the total amount of zinc in the cleft as $[Zn]_T$ and to the concentration of the unbound or free zinc as $[Zn]^{2+}$. These concentrations may increase due to the release of zinc from the glutamate vesicles when a signal arrives at the presynaptic area and the vesicles open up and pour their contents into the cleft. They may also decrease due to the uptake of zinc into the presynaptic region where it will eventually find its way back into the glutamate vesicles, thus closing the cycle. Therefore, the rate of change of the total concentration of zinc in the cleft is given by

$$\frac{d[Zn]_T}{dt} = R(t) - U(t) \tag{1}$$

where $R(t)$ represents the release, i.e., the rate at which zinc is released from the presynaptic area and enters the cleft, and $U(t)$ represents the uptake, i.e., the rate at which zinc leaves the cleft and is reabsorbed by the presynaptic region.

As zinc and glutamate are assumed to be released simultaneously, it is reasonable to expect the rates of change of their concentrations to follow the same pattern. Therefore, it is assumed that the function $R(t)$ will have a fast rising phase, followed by a slow decay, as is known to happen for glutamate. In this study, R(t) and U(t) are described by alpha functions:

$$R(t) = A_1 t\, e^{-\frac{t}{\tau_1}} \tag{2}$$

where A_1 and τ_1 are constant values that define the height and position of the peak of the release function, and

$$U(t) = A_2 t e^{-\frac{t}{\tau_2}} \tag{3}$$

where A_2 and τ_2 are again constant values. We choose τ_2 to be much larger than τ_1, as it is well-known that the uptake is usually much slower than the release. These four constants cannot be all independent from each other, as the total concentration of zinc must go back to its resting value so that equilibrium is reached again. A_1, τ_1, and τ_2 are chosen to be the independent parameters and A_2 to depend on them.

Eq. (1) is easily integrated, yielding

$$[Zn]_T = [Zn]_{T_r} + A_1\tau_1\left[\frac{\tau_1}{\tau_2}(t+\tau_2)e^{-t/\tau_2} - (t+\tau_1)e^{-t/\tau_1}\right] \tag{4}$$

where $[Zn]_{T_r}$ denotes the resting value of the total zinc concentration and

$$A_2 = A_1\left(\frac{\tau_1}{\tau_2}\right)^2, \tag{5}$$

so as to ensure that $[Zn]_T$ will go back to the basal value $[Zn]_{T_r}$, as $t \to \infty$.

Once zinc is released into the cleft, it will react with several different sites, denoted here by X_i, $i = 1, \ldots, N$, thus forming ZnX_i complexes. It is assumed that all complexes have 1:1 stoichiometry so that the reaction that takes place can be written as

$$Zn^{2+} + X_i \leftrightarrow ZnX_i, i = 1, \ldots, N. \quad (6)$$

Naturally, the total amount of zinc in the cleft must be equal to the sum of the amount of unbound zinc to the amount of zinc that is bound to all the available sites in the cleft. The total concentration of zinc in the cleft can then be written as

$$[Zn]_T = [Zn^{2+}] + \sum_{i=1}^{N}[ZnX_i], \quad (7)$$

where $[ZnX_i]$ represents the concentration of the complex ZnX_i and the sum extends to all the sites to which zinc can bind.

A similar reasoning can be applied to each binding site X_i. Let $[X_i]_T$ denote the total concentration of X_i in the cleft and $[X_i]$ the concentration of the free site X_i, that is, the concentration of X_i that is not bound to zinc. Then, assuming that there are no other ions competing with zinc for binding to X_i, one must clearly have

$$[X_i]_T = [X_i] + [ZnX_i], i = 1, \ldots, N. \quad (8)$$

The differential equation that describes the dynamics of $[ZnX_i]$ can be obtained from the reaction shown in Eq. (6). Defining $k_{on,i}$ and $k_{off,i}$ as the association and dissociation rate constants for the reactions involving the site X_i and zinc, this equation becomes

$$\frac{d}{dt}[ZnX_i] = k_{on,i}[Zn^{2+}][X_i] - k_{off,i}[ZnX_i] \qquad i = 1, \ldots, N. \quad (9)$$

Eq. (8) can be used to remove the variable $[X_i]$ from Eq. (9) and obtain

$$\frac{d}{dt}[ZnX_i] = k_{on,i}[Zn^{2+}][X_i]_T - [ZnX_i](k_{on,i}[Zn^{2+}] + k_{off,i}) \quad i = 1, \ldots, N. \quad (10)$$

Apparently, this equation represents a system of linear, uncoupled, first-order differential equations, which could easily be solved numerically. However, the situation is a little more complicated than that for two reasons.

The first reason is that two parameters required in the differential equation, namely, $k_{on,i}$ and $k_{off,i}$, are not known for the reactions of zinc with some of the sites. This issue can be circumvented by assuming that, for those cases, the reactions are so fast, compared to the others, that they quickly adapt to the changes of $[Zn^{2+}]$. This is done so that the value of $[ZnX_i]$ at any given time is similar to its equilibrium value $[ZnX_i]_{eq}$ for that particular free zinc concentration $[Zn^{2+}]$, which is given by

$$\frac{d}{dt}[ZnX_i]_{eq} = 0 \Rightarrow [ZnX_i]_{eq} = \frac{k_{on,i}[Zn^{2+}][X_i]_T}{k_{on,i}[Zn^{2+}] + k_{off,i}}, \quad (11)$$

yielding

$$[ZnX_i]_{eq} = \frac{[Zn^{2+}][X_i]_T}{[Zn^{2+}] + k_{D,i}} \quad (12)$$

where the dissociation constant, $k_{D,i}$, is given by

$$k_{D,i} = \frac{k_{off,i}}{k_{on,i}}. \tag{13}$$

Thus, if there are known values for $k_{D,i}$ (or IC_{50} or EC_{50}) for all the reactions in question, the concentrations $[ZnX_i]$ can be determined using Eq. (12).

The second issue mentioned above is the presence of $[Zn^{2+}]$ in the equations. The concentration of free zinc is an unknown function of time, which will obviously depend on the amount of zinc that is bound to all the sites X_i, so it will depend on the concentrations of all the complexes $[ZnX_i]$ through Eq. (7), which can be written as

$$[Zn^{2+}] = [Zn]_T - \sum_{i=1}^{N}[ZnX_i]. \tag{14}$$

In the model presented here, the total concentration of zinc is a known function of time, so Eq. (14) will couple all the differential Eq. (10) that describe the dynamics of the zinc complexes, as well as the algebraic Eq. (12). This is easy to understand. If more zinc couples to site X_i, there will be less free zinc left to couple to the other sites.

It should be noticed that this model imposes a certain amount of total zinc in the cleft at any specified time. When choosing the parameters of the alpha functions, care must be taken in order not to allow the total amount of zinc in the cleft to be less than the amount of zinc that is bound to the sites X_i at any time.

2.2 Numerical calculations

The task is now to solve the differential Eqs. (10) and the algebraic Eqs. (12), where each one of these equations is coupled to all the others by Eq. (14).

The differential Eqs. (10) can be solved by standard methods, which shall now be described [53]. Let, for the sake of simplicity, the concentrations of the complexes be denoted by $f_i(t)$ and the right-hand side of Eq. (10) be denoted by $F_i(t, f_1, \ldots, f_N)$. It should be noted that the concentration of free zinc in the right-hand side of that equation will depend on time and on all the complex concentrations, which means that F_i will have to depend also on time and on all the complex concentrations. Eq. (10) will then take the form

$$\frac{df_i}{dt} = F_i(t, f_1, \ldots, f_N), i = 1, \ldots, N. \tag{15}$$

The goal is to integrate this set of equations from $t = 0$ until $t = T$ for some final time T. In order to achieve that goal, the total time interval T is divided into n equal small time intervals Δt so that

$$\Delta t = \frac{T}{n}. \tag{16}$$

The initial values $f_i(0)$ are assumed to be known. Then, the calculation follows step by step, starting with the evaluation of the values of f_i at time Δt, next the values at time $2\Delta t$, and so on, until the final time $T = n\Delta t$ is reached and the final values are computed. In order to determine the values at time $(k+1)\Delta t$, where k is an integer, assuming that the values at time $k\Delta t$ are known, an expansion of $f_i((k+1)\Delta t)$ in a Taylor series is performed:

$$f_i((k+1)\Delta t) = f_i(k\Delta t) + \left[\frac{df_i}{dt}\right]_{t=k\Delta t} \Delta t + O(\Delta t^2), \tag{17}$$

where Eq. (15) can be used to yield

$$f_i((k+1)\Delta t) = f_i(k\Delta t) + F_i(t, f_1(k\Delta t), \dots, f_N(k\Delta t))\Delta t + O(\Delta t^2). \tag{18}$$

In this equation, $F_i(t, f_1, \dots, f_N)$ is evaluated at time $t = k\Delta t$ with the values of $f_1, \dots, f_N$ also taken at time $t = k\Delta t$: $f_1(k\Delta t), \dots, f_N(k\Delta t)$. Neglecting the term of second order in Δt, Eq. (18) becomes

$$f_i((k+1)\Delta t) \cong f_i(k\Delta t) + F_i(t, f_1(k\Delta t), \dots, f_N(k\Delta t))\Delta t. \tag{19}$$

The error in this result is, of course, of second order in Δt.

In order to reach time $t = T = n\Delta t$, it is necessary to take n time steps of length Δt each. The error in the final result will be the sum of n errors of order Δt^2, which makes it of order $O(n\Delta t^2) = O(\Delta t)$, as n is of order $\frac{1}{\Delta t}$, as shown in Eq. (16). Therefore, the error can be made smaller by decreasing the value of Δt, which is equivalent to increasing the value of n. This method is called the Euler method, and the fact that the final error is of first order in Δt makes it a first-order method. It is possible to improve this method by using a more complex calculation at each time step.

In these calculations, the very popular fourth-order Runge–Kutta method was used. In one dimension (only one differential equation), it consists of defining, at time step k

$$g_1 = F(k\Delta t, f(k\Delta t))\Delta t \tag{20}$$

$$g_2 = F\left(k\Delta t + \frac{\Delta t}{2}, f(k\Delta t) + \frac{g_1}{2}\right)\Delta t \tag{21}$$

$$g_3 = F\left(k\Delta t + \frac{\Delta t}{2}, f(k\Delta t) + \frac{g_2}{2}\right)\Delta t \tag{22}$$

$$g_4 = F(k\Delta t + \Delta t, f(k\Delta t) + g_3)\Delta t. \tag{23}$$

Then, the value of $f((k+1)\Delta t)$ at the next time step will be given by

$$f((k+1)\Delta t) = f(k\Delta t) + \frac{g_1}{6} + \frac{g_2}{3} + \frac{g_3}{3} + \frac{g_4}{6} + O(\Delta t^5). \tag{24}$$

This result can be easily checked by following the lengthy process of expanding all terms in a Taylor series.

After summing all the n increments, the final result will have an error of order Δt^4, thus converging much faster than the Euler method. It is easy to extend this result to several dimensions as in the present case.

Usually, the routines that use this method make some sort of quality control of the intermediate results [53]. This can be achieved in several ways, like by comparing a fourth-order result with a fifth-order result. The difference between the two should provide a reasonable estimate for the error. If the error is too small, the routine increases the value of Δt, in order for the calculation to proceed at a faster pace. If the error is too large, then the result is rejected, and a new attempt is made with a smaller step.

This method provides the values of the complex concentrations whose dynamics is described by Eq. (10) after one time step. However, it is also necessary to calculate the values of the other complex concentrations and of the free zinc concentration. This can be achieved by iterating Eqs. (12) and (14) in the following way: using the results obtained by the Runge–Kutta method for the reactions for which the rate constants are known, together with the previous values for the results obtained for the other complex concentrations in Eq. (14), an approximate value for the free zinc concentration is obtained. This value can then be used in Eq. (12), providing new values for the complex concentrations whose rate constants are not known and which are assumed to be in equilibrium. Then this procedure is repeated using these new results in Eq. (14) to obtain again a new value for the free zinc concentration, which is used once more in Eq. (12). The process is repeated until self-consistency is obtained, i.e., until the change of the concentration values after one iteration is negligible.

3. Results

The data indicated in **Table 1** allowed a variety of estimates of total, free, and complexed zinc changes at the hippocampal mossy fiber synapses from CA3 area. The studies considered the release processes in a synaptic cleft, both from a single and from multiple vesicles, and also the corresponding uptake. **Table 2** shows the data that was used to describe a single vesicle event and two possible multiple vesicle events which are denoted by short and long, where the latter corresponds to a longer time before the maximum concentration of free zinc is obtained, as well as a larger value for that maximum. **Table 2** includes also the values of the parameters

Zinc binding molecules	Concentration		k_{on}	k_{off}	K_D	IC_{50} (EC_{50})
	Molecules / μm^2	µM	$M^{-1}s^{-1}$	s^{-1}	µM	µM
NMDA NR1a-NR2A	18 [29]	0.042	10^8 [28]	0.6 [28]	0.006 [28]	-----
NMDA NR1a-NR2B	18 [29]	0.042	$3x10^7$	15 [28]	-----	0.49 [28]
ATP (free)	-----	0.003 [53]	-----	-----	6 [54]	-----
GLAST (EAAT1)	2300 [30]	5.3 [30]	-----	-----	-----	9.9 [31]
K_{ATP}	-----	0.111[35]	-----	-----	-----	20 [36]
EAAT4	58 [37,38]	0.135	-----	-----	-----	50 [39]
AMPA receptor	200 [32]	0.465	-----	-----	-----	(23) [33]
KA receptor	50 [29]	0.116	-----	-----	-----	67 [34]
N-VDCCs	196 [40]	0.456	-----	-----	-----	69 [41]
L-VDCCs	-----	$2x10^{-6}$ [42]	-----	-----	-----	69 [41]

See text for the conversion of the concentrations from molecules / μm^2 to µM. In the last column all numbers correspond to IC_{50} values except for the AMPA receptor, where the EC_{50} value is presented.

Table 1.
Concentrations of the sites and rate and dissociation constants, IC_{50} and EC_{50}, for the binding reactions [54, 55]. Source: Reproduced from Quinta-Ferreira et al. [12], with permission from Springer Nature.

			Single	Short	Long
Free Zinc	Basal [Zn^{2+}]		1 nM	1 nM	1 nM
	Maximum [Zn^{2+}]		10 nM	100 nM	1 μM
	Time to peak		1 ms	10 ms	100 ms
Alpha Function	R(t)	A_1	1.55 Ms^{-2}	9.25 x 10^{-2} Ms^{-2}	1.32 x 10^{-2} Ms^{-2}
		τ_1	9.59 x 10^{-5} s	1.31 x 10^{-3} s	1.10 x 10^{-2} s
	U(t)	A_2	1.76 x 10^{-8} Ms^{-2}	3.99 x 10^{-8} s	1.11 x 10^{-6} Ms^{-2}
		τ_2	0.90 s	2.00 s	1.20 s

Table 2.
Assumed values of free zinc and model parameters for the release (R(t)) and uptake (U(t)) functions, for the single, short, and long zinc release processes. Source: Adapted from Quinta-Ferreira et al. [12], with permission from Springer Nature.

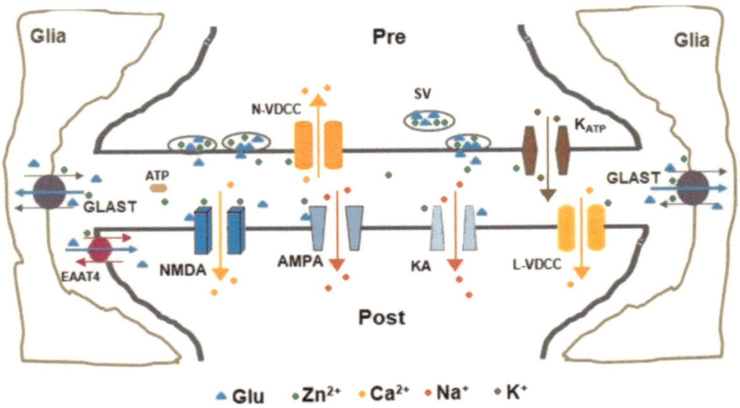

Figure 1.
Mossy fiber synaptic components from hippocampal CA3 area. Diagram of the mossy fiber synapse showing synaptic vesicles (SV), glutamate transporters (GLAST and EAAT4), glutamate receptors (NMDA, AMPA, and KA), voltage-dependent calcium channels (N- and L-VDCCs), ATP-sensitive potassium channels (K_{ATP}), and ATP molecules. The different ions (Zn^{2+}, Mg^{2+}, Na^{+}, and K^{+}) are represented by dots and the neurotransmitter glutamate (Glu) by filled triangles. Reproduced from Quinta-Ferreira et al. [12], with permission from Springer Nature.

for the release, $R(t)$, and uptake, $U(t)$, alpha functions defined in Eqs. (2) and (3), for those three cases.

Figure 1 represents the major cellular mechanisms (ionic channels (N-VDCCs, K_{ATP} and L-VDCCs), glutamate receptors (AMPA, NMDA and KA), and glutamate transporters (GLAST, EAAT4)) forming complexes with cleft free zinc. In this work the high- (NR1a-NR2A) and low (NR1a-NR2B)-affinity NMDA receptor sites will simply be mentioned as NMDA 2A and NMDA 2B, respectively.

Since the curves corresponding to the short release process have similar time courses to those of the long release events, as can be seen in **Figure 4**, only the smaller and faster (single process) and the larger and slower (long process) curves are shown in **Figures 2** and **3**. Thus, in **Figure 2**, the upper panels represent superimposed signals of the time derivative of the total zinc, dZn_T/dt; the total zinc, Zn_T; and the free zinc, Zn^{2+}, concentrations. Except for the first panel, all the other panels are represented on a semilog scale due to the very large range of the signal amplitudes of the data displayed in **Figure 2**. This allows for an easier comparison of the initial and maximum concentrations and also of the time courses of the formed complexes. The changes in the dZn_T/dt curves occur very rapidly and are over

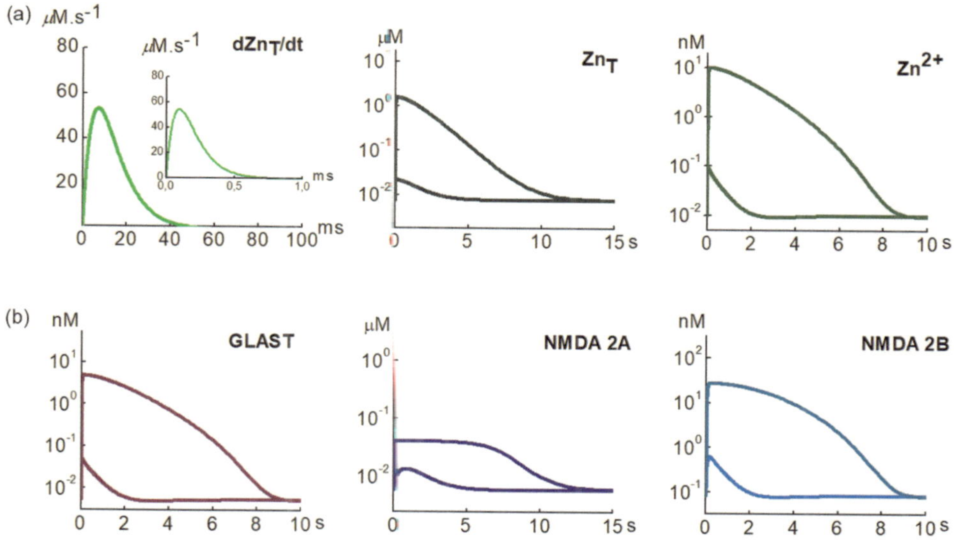

Figure 2.
Modeled zinc changes associated with single and long release processes at hippocampal mossy fiber synapses. (a) The traces represent dZn_T/dt (left), which is given by the difference between the release and the uptake functions, the total (center), and the free (right) zinc concentrations as a function of time. The inset on the left panel shows dZn_T/dt for the single event, in an expanded time scale. (b) Zinc complexes formed with the GLAST (left) glutamate transporter sites and the NMDA 2A (center) and NMDA 2B (right) glutamate receptor sites. In each panel, the smaller and the larger traces are for the single and long release processes, respectively. Note the different scales.

within 1 and 10 ms, for the single and long processes, respectively. The same curves have maxima at about 100 μs and 1 ms measuring both approximately 55 $\mu M\ s^{-1}$. These curves (see Eqs. (1), (2), and (3)), were obtained subtracting the uptake, U(t), from the release, R(t), alpha functions.

The lower part of **Figure 2** shows again superimposed signals associated with the single and long release processes but for the GLAST, NMDA 2A and NMDA 2B complexes.

In these findings, the peak amplitude of free zinc is about one-half of the peak amplitude of total zinc for the single release process and about two thirds for the long process. Most total and free zinc changes occur within about 5 s, for the single stimulation, and about 15 s, for the long one (**Figure 2**). The major zinc-binding sites considered in the model are listed in **Table 1**, where the reaction rate constants and the dissociation constants (or the IC_{50} or EC_{50} values), if available from previous works, are also indicated.

The NMDA 2A and NMDA 2B curves, with peaks in the nM range, were built using reaction rate constants. In both cases, the time course for the long process is slower than the time course for the single process (**Figure 2**).

All the other complexes, with unknown rate constants, are considered to be always in equilibrium with free zinc (see Section 2) and have, thus, identical decreasing shapes to that of free zinc. The dynamics of these complexes, namely, AMPA, N-type VDCCs, K_{ATP}, EAAT4, KA, and ATP, listed in decreasing order of magnitude, happens in less than 5 s (single process) or 10 s (long process), depending on the intensity of stimulation (**Figures 2** and **3**). It is interesting to note the faster recovery of the NMDA 2B complex, with respect to NMDA 2A, which is characterized by higher-affinity zinc binding and thus slower dissociation rate, for both the single and long processes.

In the model, based on experimental results, these processes are characterized by an initial free zinc value, $[Zn^{2+}]$, of 1 nM, and by maximum and time-to-peak

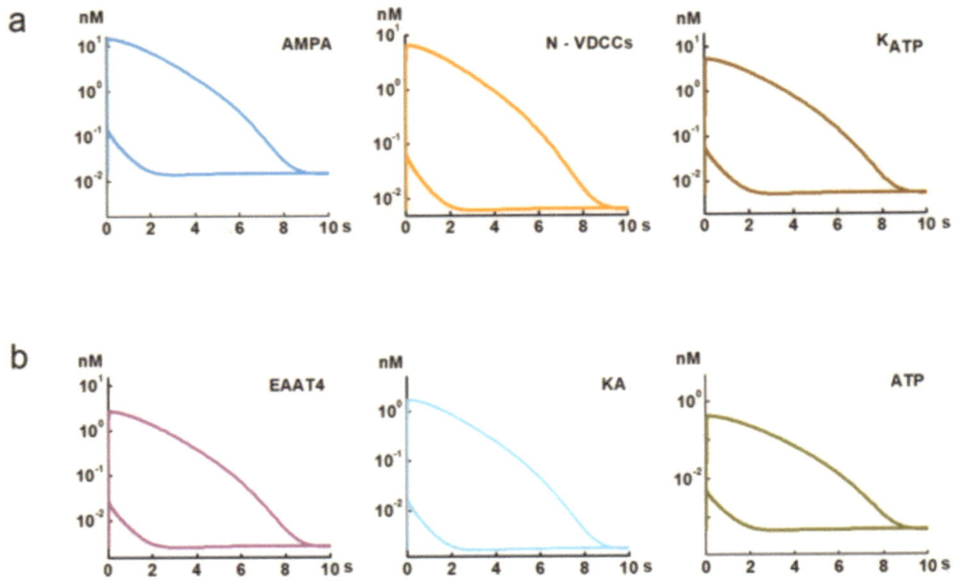

Figure 3.
Zinc complexes for the single and long release processes with time courses similar to those of free zinc. Complexes with unknown on and off-rate constants, assumed to be always at equilibrium with free zinc, formed with (a) AMPA receptors (left), N-VDCCs (center), and K_{ATP} channels (right). (b) The same type of complexes formed with EAAT4 transporters (left), KA receptors (center), and ATP molecules (right). For both the single (smaller traces) and the long (larger traces) release processes, the signals are displayed by decreasing order of amplitude. Note the different scales.

values of 10 nM and 1 ms for the single stimulation and 1 μM and 100 ms for the long release process. In the single case (and also in the short process), zinc clearance is mainly due to zinc binding to the NMDA 2A and GLAST sites, while in the long stimulation, it is essentially mediated by the formation of GLAST complexes. All the other complexes, characterized by lower affinities, are formed in smaller concentrations, as can be seen in **Figures 2** and **3**.

An overview of the main results of this study can be observed in **Figure 4**, for the single, short, and long processes.

4. Discussion

Mathematical models are a highly valuable tool for the study of synaptic zinc dynamics and in particular of cleft zinc changes. After release, zinc interacts with a variety of pre- and postsynaptic mechanisms which, together with uptake, mediate cleft zinc clearance. In the simpler case, which assumes no postsynaptic zinc entry as considered in this study, all released zinc returns, after some time, to the presynaptic area.

Previous works have suggested that there is no zinc entry in the postsynaptic region for concentrations below 10 μM [18, 20, 27, 56]. As previously reported [12], our model assumes different zinc release events that lead to maximum cleft free zinc concentrations in the range 10 nM to 1 μM, which are below 10 μM. With these values, there should be no zinc entering to the postsynaptic area.

The released zinc can form complexes with various synaptic zinc-binding sites existing mainly on pre- and postsynaptic VDCCs (N- and L-types), K_{ATP} channels, ionotropic glutamate receptors (AMPA, KA, and NMDA), and also cleft free molecules (ATP) and glial glutamate transporters (EAAT4). The mathematical model

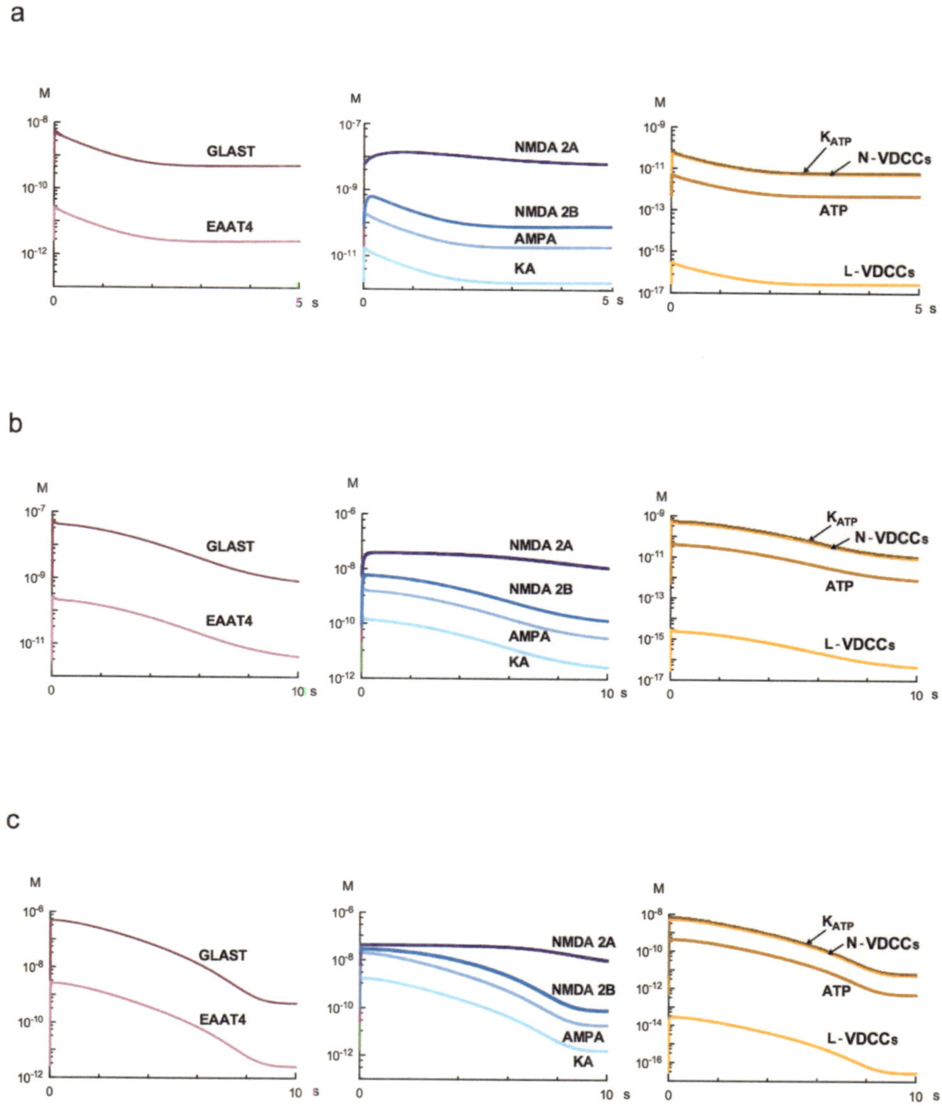

Figure 4.
Comparison of the amplitude and time course of the complexes associated with the single, short, and long processes. (a) Superimposed traces of the zinc complexes formed with GLAST and EAAT4 transporters (left), NMDA 2A, NMDA 2B, AMPA, and kainate receptors (center) with K_{ATP} and N- and L-type VDCCs (right) evoked by a single release event. The latter panel includes also zinc binding to ATP molecules (b) and (c). Similar to (a), but for short (b) and long (c) release processes. Reproduced from Quinta-Ferreira et al. [12], with permission from Springer Nature.

considered in this study was designed to obtain estimates of the dynamic behavior of the zinc complexes formed with the mentioned mechanisms [12], assuming previously reported resting and site zinc concentrations and binding constants, which are summarized in **Table 1**. The formation of the complexes was triggered by three assumed types of stimuli, corresponding to single, short, and long release events, with the maximum and time to peak values, based also on existing experimental findings, indicated in **Table 2**. In most cases, where the on and off-rate constants were not known, the dissociation constant K_D (or the corresponding IC_{50} or EC_{50}) values were considered in the equations. In these cases the complexes were considered to attain very rapidly the equilibrium concentration, following each free zinc concentration change. This assumption was made for the following reasons: the steps involved in the formation of a complex include the movement of a zinc ion toward the site, the binding, and the dissociation from it, the latter occurring after a conformational change of the bound target. The value of the on-rate constant is mainly determined by diffusion and binding. The speed of a reaction is largely determined by the diffusion value, considered for all ions as 10^9–10^{10} s^{-1} [57] and, to some extent, by the rate with which water leaves the ionic solvation sphere. For zinc, this rate and the on-rate constant are considered to be above 10^7 s^{-1} and 10^8 M^{-1} s^{-1}, respectively [57]. We can now introduce values of on-rate constants between 10^7 and 10^8 M^{-1} s^{-1}, in agreement with the values for the NMDA sites used in the model and the dissociation, IC_{50} and EC_{50}, values in **Table 1**. In these cases, the off-rate constants have values in the range 60 s^{-1} to approximately 7000 s^{-1}, which exceed, by two to four orders of magnitude, the off-rate constant (0.6 s^{-1}) for the NMDA 2A reaction. Thus, since for those complexes unbinding occurs very rapidly, the assumption that they are always in equilibrium with free zinc is justified.

The computational study led to the representation of a variety of time-varying curves to illustrate, for different stimuli, the release minus uptake and the concentrations of total and free zinc, as well as of the various zinc complexes formed. Assuming, in a single release process, a low level of stimulation, causing free zinc to reach 10 nM in the cleft, the predominant complexes are formed with the high-affinity NMDA 2A glutamate receptor sites and with the lower-affinity and highly concentrated GLAST glutamate transporters from glial cells [58]. The concentrations of the other complexes are lower by several orders of magnitude. As mentioned before, all ligands for which the reaction rate constants are not known are assumed to be in equilibrium with free zinc. For this reason all these complexes, and also those associated with the short and long release events, have a similar time course to that of free zinc.

If a 10 times higher free zinc concentration, 100 nM, is attained in the cleft, following the more intense short stimulation, the most abundant zinc complex is now the GLAST one, followed by the NMDA 2A which lasts longer than any of the other complexes. This is due to the much higher affinity of the NMDA 2A sites for zinc. Similar properties apply to the third type of estimated curves, associated with the existence of 1 μM free zinc in the cleft, produced by a longer stimulus. The main differences, with respect to the short stimulus situation, are that the concentration of the NMDA 2A complex remains high for a longer period and all formed complexes have larger amplitudes than in the case of the short stimulus. The fact that externally applied zinc (100 nM) was found to inhibit postsynaptic NMDA currents, at hippocampal CA3 neurons [27], is in agreement with the idea that zinc binds to and inhibits the NMDA receptors. The remaining signals, with much smaller amplitudes, have thus minor or negligible roles in accounting for cleft zinc removal.

The most intense stimulation considered in this study, which assumed a maximum zinc cleft concentration of 1 μM, reveals that, as observed in the short case, GLAST is the complex formed in higher concentration. It is followed by the NMDA 2A complex that saturates when all the corresponding zinc sites (about 40 nM) are bound to zinc. All the other bindings have again much smaller contributions to zinc clearance, especially the complex formed with L-type VDCCs, with a concentration in the fM range.

In conclusion, for a single stimulus, the NMDA 2A high-affinity sites are the most involved in the initial clearance process, while for the stronger stimulations considered (short and long), this role is taken by the highly abundant GLAST complexes. In all cases, uptake has a much slower time course.

At hippocampal CA1 synapses, a single release event may be associated with the release of up to 10 vesicles [58]. In particular, an individual mossy fiber axon has approximately 15 giant boutons (3–5 μm diameter at 150 μm intervals) and 37 release sites (active zones) each [51, 52, 59]. Single boutons contain about 20 active zones at 0.45 μm intervals and 16,000 vesicles, being 1400 ready for release that may be multivesicular, thus leading to very intense release processes [51, 60]. It has been reported that the free or loosely bound zinc and glutamate vesicular concentrations are 1–5 and 60–210 mM, respectively [16, 27, 47]. Let's assume that zinc and glutamate, which have similar diffusion coefficients [20, 27], are co-released in the same proportion and that, as what happens for glutamate, the cleft released zinc concentration becomes uniform in tens of microseconds [16]. It can be estimated that the volume of a mossy fiber-CA3 cleft (20 nm width, 3 μm radius) is much higher (approximately 17,000 times) than that of a vesicle (40 nm diameter) [27, 49–51, 61, 62]. The cleft real volume is significantly smaller, by about 25%, if the volume of densely packed conic (20 nm height) dendritic protrusions is subtracted [63]. In this case the discharge of an individual vesicle would lead to the following initial concentrations: 80–400 nM for zinc and 4–20 μM for glutamate. There is a large difference between the radii and, thus, the volumes of the mossy fiber-CA3 and the CA3-CA1 clefts since the latter has a 20 nm width and only 250 nm radius. As a consequence, in the CA1 region, the cleft glutamate range of concentrations is estimated to be around 0.4–2 mM and is thus close to the previously reported ranges of values, 1–5 mM [16] and 0.25–11 mM following an individual vesicle release [64]. Another important issue is the role of the connection between the cleft and extrasynaptic regions with much higher volume [65]. This volume has to be added to that of the cleft, since it forms a large part of the space where the zinc concentration changes occur. A similar fact has been considered for cleft glutamate clearance in neurons of the central nervous system, where the glutamate concentration decreased very rapidly (1–5 ms) 100–500 times [65]. Let us assume again that zinc and glutamate diffuse in a similar way [20, 27]. In this case the concentration range of cleft free zinc changes, evoked by an individual vesicle discharge and after diffusing away from the cleft, will be 0.8–4 nM. If multivesicular release occurs [60], the amount of zinc in the cleft will be significantly larger.

Another estimate of cleft zinc discharge and uptake can be obtained from fluorescent glutamate signals associated with single or repetitive stimulation applied to cultured hippocampal neurons [66]. These authors have found that for the single and short types of stimulation, the maximum concentrations of glutamate were around 0.3 and 0.8 μM, respectively, occurring clearance in less than 1 s, for the single, and 2 s, for the short stimuli. If, as previously estimated, there is about 50 times more glutamate in the cleft, the equivalent maximum concentration range for free zinc will be 6–16 nM.

Previous work has reported that the resting free zinc concentration in the cleft is below 10 nM, meaning that the NMDA 2A sites, which are highly sensitive for zinc, will not become saturated by zinc, and also that the amount of zinc discharged by an individual stimulus does not seem to alter much postsynaptic NMDA currents [27]. Our estimates, for more intense stimulations such as the short and long processes, which are associated with 10 and 100 times more free zinc in the cleft, respectively, than for the single process, indicate that the NMDA 2A sites (K_D = 6 nM) are nearly or fully occupied as reported earlier [10, 27, 28]. For the short and long stimulations, the amount of GLAST complexes is approximately 2 and 10 times more than for NMDA 2A complexes, respectively. Also, only for these stronger stimulation protocols, the much lower-affinity NMDA 2B and AMPA receptors become significantly occupied. The K_{ATP} channels and the EAAT4 transporters form a reasonable amount of zinc complexes in spite of their smaller affinity, because their concentrations are higher than those of the NMDA sites. As expected, the zinc complexes with lower concentrations and association rates, such as those formed with the kainate receptors and ATP molecules, are only significant following the most intense (long) stimulation. It should also be noticed that the maximum cleft zinc concentration included in the model, 1 μM, does not lead to the inhibition of N- or L-type VDCCs by zinc, since the threshold, half, and almost full blockade concentrations are <5, 69, and 150–200 μM, respectively [40, 42]. For all protocols considered, very small concentrations (in the order of pM-fM) of the VDCC complexes are formed.

The evaluation of the dynamics of synaptic zinc complexes considered in this work contributes to a wider knowledge about synaptic zinc changes. Identifying the main zinc mechanisms involved in mossy fiber zinc clearance is of major importance, considering the potential protective or toxic roles of released zinc at these highly excitable synapses.

Acknowledgements

M.E. Quinta-Ferreira is grateful to Dr. C.C.A.M. Gielen for providing the conditions to start this work. It was funded by the strategic project UID/NEU/04539/2013.

Author details

Johnattan C.S. Freitas[1], João N. Miraldo[1], Carlos Manuel M. Matias[1,2,3]*, Fernando D.S. Sampaio dos Aidos[2,4], Paulo J. Mendes[1,5], José C. Dionísio[6], Rosa M. Santos[2,7], Luís M. Rosário[2,7], Rosa M. Quinta-Ferreira[8] and Emília Quinta-Ferreira[1,2]

1 Department of Physics, UTAD, University of Coimbra, Coimbra, Portugal

2 Center for Neuroscience and Cell Biology (CNC), University of Coimbra, Coimbra, Portugal

3 University of Trás-os-montes and Alto Douro (UTAD), Vila Real, Portugal

4 Department of Physics, CFisUC, University of Coimbra, Coimbra, Portugal

5 Laboratory of Instrumentation and Experimental Particles Physics (LIP), Coimbra, Portugal

6 Department of Animal Biology, University of Lisbon, Lisbon, Portugal

7 Department of Life Sciences, University of Coimbra, Coimbra, Portugal

8 Department of Chemical Engineering, CIEPQPF-Research Centre of Chemical Process Engineering and Forest Products, University of Coimbra, Coimbra, Portugal

*Address all correspondence to: cmatias@utad.pt

References

[1] Frederickson C. Neurobiology of zinc and zinc-containing neurons. International Review of Neurobiology. 1989;**31**:145-238

[2] Vallee B, Falchuk K. The biochemical basis of zinc physiology. Physiological Reviews. 1993;**73**:79-118

[3] Huang E. Metal ions and synaptic transmission: Think zinc. Proceedings of the National Academy of Sciences of the United States of America. 1997;**94**: 13386-13387

[4] Frederickson CJ, Koh JY, Bush AI. The neurobiology of zinc in health and disease. Nature Reviews. Neuroscience. 2005;**6**:449-462

[5] Perez-Clausell J, Danscher G. Intravesicular localization of zinc in rat telencephalic boutons: A histochemical study. Brain Research. 1985;**337**:91-98

[6] Harrison NL, Gibbons SJ. Zn^{2+}: An endogenous modulator of ligand- and voltage-gated ion channels. Neuropharmacology. 1994;**33**:935-952

[7] Smart T, Xie X, Krishek B. Modulation of inhibitory and excitatory amino acid receptor ion channels by zinc. Progress in Neurobiology. 1994;**42**: 393-441

[8] Lin D, Cohen A, Coulter D. Zinc-induced augmentation of excitatory synaptic currents and glutamate receptor responses in hippocampal CA3 neurons. Journal of Neurophysiology. 2001;**85**:1185-1196

[9] Ruiz A, Walker M, Fabian-Fine R, Kullmann D. Endogenous zinc inhibits GABA (A) receptors in a hippocampal pathway. Journal of Neurophysiology. 2004;**91**:1091-1096

[10] Erreger K, Traynellis SF. Allosteric interaction between zinc and glutamate binding domains on NR2A causes desensitization of NMDA receptors. The Journal of Physiology. 2005;**569**:381-393

[11] Paoletti P, Vergnano A, Barbour A, Review CM. Zinc at glutamatergic synapses. Neuroscience. 2009;**158**: 126-136

[12] Quinta-Ferreira ME, Sampaio dos Aidos FDS, Matias CM, Mendes PJ, Dionísio JC, Santos RM, et al. Modelling zinc changes at the hippocampal mossy fiber synaptic cleft. Journal of Computational Neuroscience. 2016;**41**: 323-337

[13] Clements J, Lester R, Tong G, Jahr C, Westbrook G. The time course of glutamate in the synaptic cleft. Science. 1992;**258**:1498-1501

[14] Yamada W, Zucker R. Time course of transmitter release calculated from stimulations of a calcium diffusion model. Biophysical Journal. 1992;**61**:671-682

[15] Südhof T. The synaptic vesicle cycle: A cascade of protein-protein interactions. Nature. 1995;**375**:645-653

[16] Clements J. Transmitter timecourse in the synaptic cleft: Its role in central synaptic function. Trends in Neurosciences. 1996;**19**:163-171

[17] Sensi S, Canzoniero L, Shen P, Howard S, Koh J, et al. Measurement of intracellular free zinc in living cortical neurons: Routes of entry. The Journal of Neuroscience. 1997;**17**:9554-9564

[18] Marin P, Israel M, Glowinski J, Premont J. Routes of zinc entry in mouse cortical neurons: Role in zinc-induced neurotoxicity. The European Journal of Neuroscience. 2000;**12**:8-18

[19] Vogt K, Mellor J, Tong G, Nicoll R. The actions of synaptically released zinc

at hippocampal mossy fiber synapses. Neuron. 2000;**26**:187-196

[20] Li Y, Hough CJ, Suh SW, Sarvey JM, Frederickson CJ. Rapid translocation of Zn^{2+} from presynaptic terminals into postsynaptic hippocampal neurons after physiological stimulation. Journal of Neurophysiology. 2001;**86**:2597-2604

[21] Li Y, Hough CJ, Frederickson CJ, Sarvey JM. Induction of mossy fiber→Ca3 long-term potentiation requires translocation of synaptically released Zn^{2+}. The Journal of Neuroscience. 2001b;**21**:8015-8025

[22] Quinta-Ferreira ME, Matias CM, Arif M, Dionísio JC. Measurement of presynaptic zinc changes in hippocampal mossy fibers. Brain Research. 2004;**1026**:1-10

[23] Qian J, Noebels JL. Visualization of transmitter release with zinc fluorescencedetection at the mouse hippocampal mossy fibre synapse. The Journal of Physiology. 2005;**566**:747-758

[24] Quinta-Ferreira ME, Matias CM. Tetanically released zinc inhibits hippocampal mossy fiber calcium, zinc and postsynaptic responses. Brain Research. 2005;**1047**:1-9

[25] Ketterman J, Li Y. Presynaptic evidence for zinc release at the mossy fiber synapse of rat hippocampus. Journal of Neuroscience Research. 2008; **86**:422-434

[26] Khan M, Goldsmith CR, Huang Z, Georgiou J, Luyben TT, Roder JC, et al. Two-photon imaging of Zn^{2+} dynamics in mossy fiber boutons of adult hippocampal slices. Proceedings of the National Academy of Sciences of the United States of America. 2014;**111**: 6786-6791

[27] Vergnano A, Rebola N, Savtchenko L, Pinheiro P, Casado M, Kieffer B, et al. Zinc dynamics and action at excitatory synapses. Neuron. 2014;**82**:1101-1114

[28] Paoletti P, Arscher P, Neyton C. High-affinity zinc inhibition of NMDA NR1–NR2A receptors. The Journal of Neuroscience. 1997;**17**:5711-5725

[29] Gao X-M, Sakai K, Roberts R, Dean B, Tamminga C. Ionotropic glutamate receptors and expression of N-methyl-D-aspartate receptor subunits in subregions of human hippocampus: Effects of schizophrenia. The American Journal of Psychiatry. 2000;**157**(7):1141-1149

[30] Lehre K, Danbolt N. The number of glutamate transporter subtype molecules at glutamatergic synapses: Chemical and stereological quantification in young adult rat brain. The Journal of Neuroscience. 1998;**18**: 8751-8757

[31] Vanenberg R, Mitrovic A, Johnston G. Molecular basis for differential inhibition of glutamate transporter subtypes by zinc ions. Molecular Pharmacology. 1998;**54**:189-196

[32] Nusser Z, Lujan R, Laube G, Roberts JD, Molnar E, Somogyi P. Cell type and pathway dependence of synaptic AMPA receptor number and variability in the hippocampus. Neuron. 1998;**21**:545-559

[33] Bresink I, Ebert B, Parsons C, Mutschler E. Zinc changes AMPA receptor properties: Results of binding studies and patch clamp recordings. Neuropharmacology. 1996;**35**:503-509

[34] Mott D, Beneviste M, Dingledine R. pH-dependent inhibition of kainite receptors by zinc. The Journal of Neuroscience. 2008;**13**:1659-1671

[35] Zini S, Tremblay Roisin M-P, Ben-Ari Y. Two binding sites for [^{3}H] glibenclamide in the rat brain. Brain Research. 1991;**542**:151-154

[36] Bloc A, Cens T, Cruz H, Dunant Y. Zinc-induced changes in ionic currents of clonal rat pancreatic-cells: Activation

of ATP-sensitive K+ channels. The Journal of Physiology. 2000;**529**:723-734

[37] Furuta A, Martin LJ, Lin CL, Dykes-Hoberg M, Rothstein JD. Cellular and synaptic localization of the neuronal glutamate transporters excitatory amino acid transporter 3 and 4. Neuroscience. 1997;**81**:1031-1042

[38] Dehnes Y, Chaudhry FA, Ullensvang K, Lehre KP, Storm-Mathisen J, Danbolt NC. The glutamate transporter EAAT4 in rat cerebellar Purkinje cells: A glutamate-gated chloride channel concentrated near the synapse in parts of the dendritic membrane facing astroglia. The Journal of Neuroscience. 1998;**18**:3606-3619

[39] Mitrovic AD, Plesko F, Vanenberg RJ. Zn^{2+} inhibits the anion conductance of the glutamate transporter EAAT4. The Journal of Biological Chemistry. 2001;**276**:26071-26076

[40] Jones O, Bernstein G, Jones E, Jugloff D, Law M, et al. N-type calcium channels in the developing rat hippocampus: Subunit, complex, and regional expression. The Journal of Neuroscience. 1997;**17**:6152-6164

[41] Büsselberg D, Michael D, Evans M, Carpenter O, Haas H. Zinc (Zn^{2+}) blocks voltage gated calcium channels in cultured rat dorsal root ganglion cells. Brain Research. 1992;**593**:77-81

[42] Hell J, Westenbroek R, Warner C, Ahlijanian M, Prystay W, et al. Identification and differential subcellular localization of the neuronal class C and class D L-type calcium channel α subunits. The Journal of Cell Biology. 1993;**123**:949-962

[43] Frederickson CJ, Suh SW, Silva D, Frederickson CJ, Thomson RB. Importance of zinc in the central nervous system: The zinc-containing neuron. The Journal of Nutrition. 2000; **130**:1471-1478

[44] Colvin RA, Fontaine CP, Laskowski M, Thomas D. Zn^{2+} transporters and Zn^{2+} homeostasis in neurons. European Journal of Pharmacology. 2003;**479**: 171-185

[45] Colvin R, Bush A, Volitakis I, Fontaine C, Thomas D, Kikuchi K, et al. Insights into Zn^{2+} homeostasis in neurons from experimental and modeling studies. American Journal of Physiology. Cell Physiology. 2008;**294**: 726-742

[46] Huang L, Tepaamorndech S. The SLC30 family of zinc transporters—A review of current understanding of their biological and pathophysiological roles. Molecular Aspects of Medicine. 2013;**34**: 548-560

[47] Sensi S, Paoletti P, Koh J, Aizenman E, Bush A, Hershfinkel M. The neurophysiology and pathology of brain zinc. The Journal of Neuroscience. 2011; **31**:16076-16085

[48] Ueno S, Tsukamoto M, Hirano T, Kikuchi K, Yamada M, et al. Mossy fiber Zn^{2+} spillover modulates heterosynaptic N-methyl-d-aspartate receptor activity in hippocampal CA3 circuits. The Journal of Cell Biology. 2002;**158**: 215-220

[49] Bourne JN, Harris KM. Dendritic spines. eLS. 2007. DOI: 10.1002/9780470015902.a0000093.pub2

[50] Savtchenko LP, Rusakov DA. The optimal height of the synaptic cleft. Proceedings of the National Academy of Sciences of the United States of America. 2008;**104**:1823-1828

[51] Rollenhagen A, Lübke JHR. The mossy fiber Bouton: The "common" or the "unique" synapse? Frontiers in Synaptic Neuroscience. 2010;**2**:2. DOI: 10.3389/fnsyn.2010.00002

[52] Chicurel ME, Harris KM. Three-dimensional analysis of the structure

and composition of CA3 branched dendritic spines and their synaptic relationships with mossy fiber boutons in the rat hippocampus. The Journal of Comparative Neurology. 1992;**325**: 169-182

[53] Press WH, Teukolsky SA, Vetterling WT, Flannery BP. Numerical Recipes in Fortran. The Art of Scientific Computing. 2nd ed. Cambridge: Cambridge University Press; 1992

[54] Melani A, Turchi A, Vannucchi M, Cipriani C, Gianfriddo M, et al. ATP extracellular concentrations are increased in the rat striatum during in vivo ischemia. Neurochemistry International. 2005;**47**:442-448

[55] Jiang L, Maret W, Vallee B. The glutathione redox couple modulates zinc transfer from metallothionein to zinc-depleted sorbitol dehydrogenase. Proceedings of the National Academy of Sciences of the United States of America. 1998;**95**:3483-3488

[56] Kay A. Evidence for chelatable zinc in the extracellular space of the hippocampus, but little evidence for synaptic release of Zn. The Journal of Neuroscience. 2003;**23**:6847-6855

[57] Fraústo da Silva JJR, Williams RJP. The Biological Chemistry of the Elements. The Inorganic Chemistry of Life. New York: Oxford University Press; 1991

[58] Conti R, LIsman J. The high variance of AMPA receptor- and NMDA receptor-mediated responses at single hippocampal synapses: Evidence for multiquantal release. Proceedings of the National Academy of Sciences of the United States of America. 2003;**100**: 4885-4890

[59] Bischofberger J, Engel D, Frotscher M, Jonas P. Timing and efficacy of transmitter release at mossy fiber synapses in the hippocampal network. European Journal of Physiology. 2006. DOI: 10.1007/s00424-006-0093-2

[60] Hallermann S, Pawlu C, Jonas P, Heckmann M. A large pool of releasable vesicles in a cortical glutamatergic synapse. Proceedings of the National Academy of Sciences of the United States of America. 2003;**100**:8975-8980

[61] Wenzel H, Cole T, Born D, Schwartzkroin P. Ultrastructural localization of zinc transporter-3 (ZnT-3) to synaptic vesicle membranes within mossy fiber boutons in the hippocampus of mouse and monkey. Proceedings of the National Academy of Sciences of the United States of America. 1997;**94**: 12676-12681

[62] Rollenhagen A, Sätzler K, Rodríguez EP, Jonas P, Frotscher M, Lübke JH. Structural determinants of transmission at large hippocampal mossy fiber synapses. The Journal of Neuroscience. 2007;**27**:10434-10444

[63] Savtchenko LP, Rusakov DA. Glutamate escape from a tortuous synaptic cleft of the hippocampal mossy fibre synapse. Neurochemistry International. 2004;**45**:479-484

[64] Harris KM, Sultan P. Variation in number, location and size of synaptic vesicles provides an anatomical basis for the nonuniform probability of release at hippocampal CA1 synapses. Neuropharmacology. 1995;**34**:1387-1395

[65] Kessler JP. Control of cleft glutamate concentration and glutamate spill-out by perisynaptic glia: Uptake and diffusion barriers. PLoS One. 2013; **8**(8):e70791

[66] Hires S, Zhu Y, Tsien R. Optical measurement of synaptic glutamate spillover and reuptake by linker optimized glutamate-sensitive fluorescent reporters. Proceedings of the National Academy of Sciences of the United States of America. 2008;**105**: 4411-4416